VOLCANIC FIREARMS

PREDECESSOR TO THE WINCHESTER RIFLE

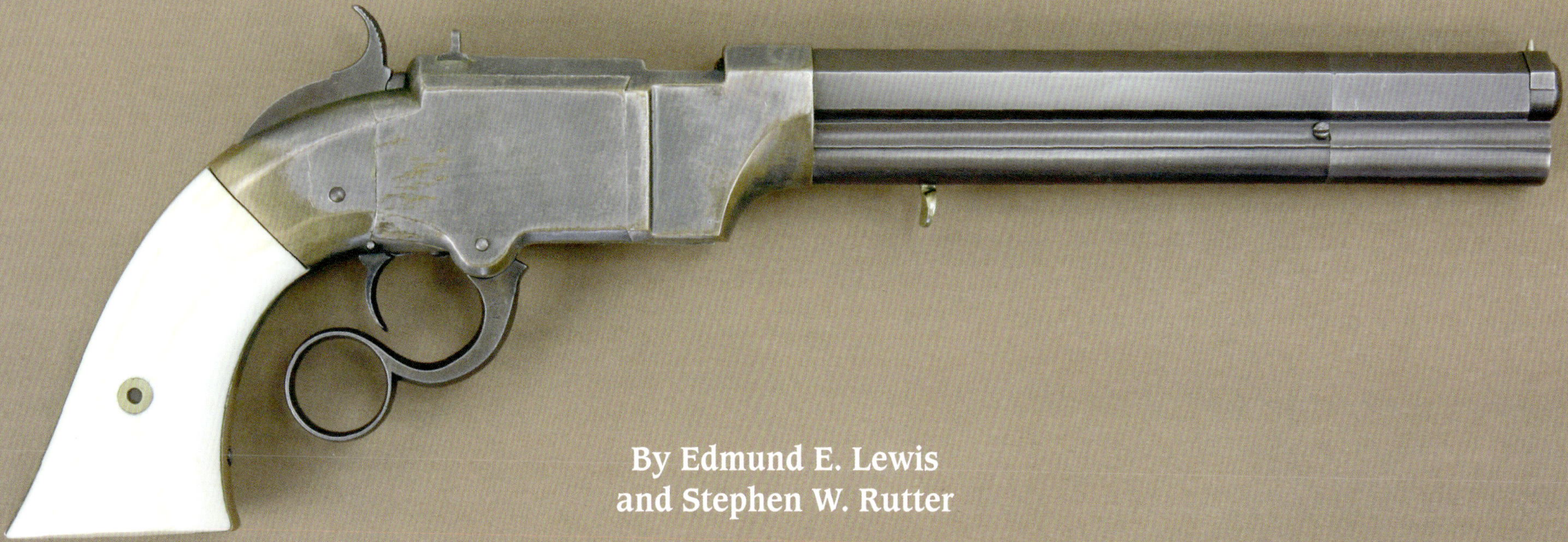

By Edmund E. Lewis
and Stephen W. Rutter

Photography by Stuart Smith and Chip Morton

Edmund E. Lewis and Stephen W. Rutter
Volcanic Firearms — Predecessor to the Winchester Rifle
Woonsocket, R.I.: ANDREW MOWBRAY INCORPORATED — PUBLISHERS
160 pages

ISBN: 1-931464-48-0

To order more copies of this book, call Mowbray Publishers at 1-800-999-4697. www.gunandswordcollector.com

Printed in China.

This book was designed and set in type by Jo-Ann Langlois.
The typeface chosen was Caxton.

ON THE COVER: Ivory-gripped Smith & Wesson small-frame 3-inch pistol, serial number 1531, and 6-inch Smith & Wesson pistol, serial number 473.
Donovan Williamson collection

1 2 3 4 5 6 7 8 9 10

To Catherine, my lovely and gifted wife, who devoted countless numbers of hours to the task of editing this book. Her advice, unending support and daily patience sustained me through this project. To my children — Madeleine, Jennifer, Ed and Sam and to the loving memory of my Mother and Father.

Edmund E. Lewis

To my best friend, my wife Vicki, and to Walt Hallstein and Doug Beneck, for their mentoring and encouragement in my education and appreciation of antique firearms.

Stephen W. Rutter

acknowledgements

The authors would like to express their gratitude to the many antique arms collectors, historians, dealers and firearm auction companies who provided invaluable assistance in the preparation of this book, especially ***Carol Watson**, whose encouragement and extensive support in the photography of many of the illustrated firearms is deeply appreciated.*

The following individuals, in alphabetical order, also deserve our special thanks:

Herbert G. Houze, noted historian, author and past curator of the Cody Firearms Museum, whose scholarship and encouragement guided us throughout the writing of this book. Herb kindly shared many documents, photographs and other material from his extensive reference library.

Roy G. Jinks, eminent historian of the Smith & Wesson Company and author of History of Smith & Wesson, shared with us his extensive knowledge of these firearms and his files on Horace Smith and Daniel Baird Wesson. Roy's illustrations, photographs and other items from his personal collection were great additions to the book.

Bruce Jordan, longtime collector and connoisseur of Winchester firearms. Bruce shared his extensive knowledge of Volcanic firearms with us and allowed us to feature many rare guns from his award-winning collection.

Rob Kassab, President of the Winchester Arms Collectors Association, whose photographic endeavors added greatly to the book.

Gerald Klaz, noted longtime collector of Volcanic firearms whose great collection was the inspiration for the authors to write this book. Many rare guns of his, never previously illustrated, are shown, along with historical details about Volcanics that have never been previously recorded.

Jo-Ann Langlois, for her handsome book design and patience in working with the authors.

Stuart Mowbray, our editor and publisher, for his guidance and expertise.

Al Muchka, Curator of the Nunnemacher Gun Collection at the Milwaukee Public Museum, for photographing and allowing us access to the outstanding Volcanic rarities in their possession.

Casey Nanz, who over several years, encouraged the writing of this book and who has a great talent for discovering new material related to Volcanic firearms, all of which he kindly shared with us.

Jack Pearson, whose logistical support, research efforts and interest in the project greatly aided in making this book possible.

Donovan Williamson, a true collector of fine Winchesters who freely shared with us his extensive collection of Volcanic firearms and ephemera. His compilation of Volcanic serial numbers, gathered over a long period of time, as well as his research files on these guns, was invaluable.

Our gratitude is also extended to the following collectors who shared their collections, photographs, information and encouragement during the writing of this book:

Raymond B. Bentley
Doug Beneck
David P. Bichrest
Richard Birkeybile
Alden S. Bradstock, III
Sean Campbell
Mike Carrick
David Condon
Peter DeRose
Leah Donnelly
Steve Elwood
John Gangel
Ray T. Giles
Paul Goodwin
Frank Graves
Walt Hallstein
Perry Hansen
Kurt House
Randy Hughes
Tina Jordan
Kirt Keller
Richard J. Labowskie
Boo Lewis
Chloe Lewis
Jack Lewis, Jr.
Carl Lind
Charles May
Warren Newman
Lisa Oakes
Patrick Quinn
Roger Quisenberry
Bobby Smith
Eric Vaule
R.L. Wilson
Dave Zarneski
Penny Zarneskie

Special thanks are also given to the following companies and institutions, all of whom made significant contributions: **Amoskeag Auction Company, Inc.,** Manchester, New Hampshire; **Cody Firearms Museum,** Buffalo Bill Historical Center, Cody, Wyoming; **Czerny's International Auction House,** Sarzana, Italy; **George Mason University,** Special Collections & Archives, Fairfax, Virginia; **James D. Julia, Inc.,** Fairfield, Maine; **Little John's Auction Service,** Orange, California; **The McCracken Research Library,** Buffalo Bill Historical Center, Cody, Wyoming; **The Milwaukee Public Museum,** Milwaukee, Wisconsin. **The National Archives,** Military and Service Pension Records, Washington, D.C.; **The Northwest Museum of Arts & Culture,** Spokane, Washington; **Rock Island Auction Company,** Moline, Illinois; **The Smithsonian Museum of American History,** Washington, D.C.; **United States Military Academy,** West Point Military Museum, West Point, New York.

And finally, our appreciation and thanks to the many other collectors and individuals who we've failed to mention. Lastly, we must stress that any and all errors of fact or omission are strictly those of the authors.

introductions

Legendary names from the past come alive in this well-researched book, as we learn of the triumphs and failures that were met in the development of the first repeating firearm using a self-contained cartridge. The story of the men who developed this concept explains their intricate relationships and also the history of the collaboration between two of America's great firearms companies: Smith & Wesson and Winchester.

The research traces all of the roles played in the long development of a successful rifle — names of legendary stature, including Walter Hunt, George Arrowsmith, Lewis Jennings, Horace Smith, Daniel B. Wesson and B. Tyler Henry. The authors provide the important roles that the financial backers such as Courtland Palmer and Oliver Winchester played in keeping alive the idea of a repeating firearm with a self-contained cartridge as well as the progression of events that led to the formation of the Winchester Repeating Arms Co. The story doesn't end without the necessary part that promoters such as J. Post, C.P. Dixon and J.W. Storrs played in bringing the new firearm to the attention of the American public.

The interested reader will find this book an excellent guide not only in understanding the development of these important firearms, but also as a collector's guide to recognizing and identifying all the variations of these rare guns.

Roy Jinks, December 2010

As the direct antecedent of the Henry Rifle, it would be natural to assume that the history and production of the arms marketed under the alliterative name "Volcanic" have been fully laid out in print. Yet, that is not the case. While a few articles have recently addressed specific aspects of their use, the basic framework recounting their manufacture remains the same as those published over half a century ago by John R. Parsons and Harold K. Williamson. Unfortunately, in the absence of concrete research a good deal of what is now accepted as being true concerning these arms is nothing more than fanciful anecdotes that have been passed down by one generation of collectors to another.

This new study at last sets the record straight. Based upon period documentation drawn from a wide variety of sources, contemporary advertising and a thorough analysis of surviving "Volcanic" firearms, Drs. Lewis and Rutter have reconstructed their origins and actual production.

In doing so, they reveal their rich history and the influence they had on the development of later repeating arms. More importantly, they do so in a manner that places the "Volcanic" within its proper context. Not only are the makers of these arms discussed, but those involved in producing their ammunition and other accessories are discussed as well. The information presented in this book detailing the introduction and manufacture of all the known models or variations thereof is based upon empirical evidence not conjecture. As a result, the work's contents can be relied upon as being accurate.

Though it took more than fifty years for the "Volcanic" to receive the attention it was properly due, the publication of Volcanic Firearms: Predecessor to the Winchester Rifle *has been well worth the wait.*

Herbert G. Houze, February 2011

An exceptional silver-plated and engraved small-frame New Haven Pocket Pistol, serial number 1268. Bruce Jordon collection

table of contents

chapter one
Hunt's Volition Repeater and the Jennings and Smith-Jennings Rifles 9
Hunt's Volition Repeater 10
The Jennings Rifle 13
Variations in Jennings Breechloaders 16
The Smith-Jennings Repeater 16
The Smith-Jennings Experimental Pistol 18

chapter two
The Smith & Wesson Lever Action Firearms 19
The First "Volcanics" 19
Origin of the Name "Volcanic" 22
General Features of the Pistols 23
The Large-Frame Smith & Wesson Repeating Pistol 25
The Small-Frame Smith & Wesson Repeating Pistol 31
A Smith & Wesson Prototype .22 Caliber Pistol 34
The Smith & Wesson Pistol-Carbines 35
The Smith & Wesson Lever Action Repeating Rifles 38
A New Smith & Wesson Partnership 40

chapter three
Volcanic Ammunition 43
The Hunt Rocket Ball 44
The Smith & Wesson Metallic Cartridge Experimentations 44
The Smith & Wesson "Volcanic" Cartridges 45
Problems with the Cartridges 47
Volcanic Cartridge Boxes 47
Afterthoughts 50

chapter four
New Beginnings— The Volcanic Repeating Arms Company 50
General Features of the Pistols 53
The Volcanic Navy Pistols 54
The Volcanic Pistol-Carbine 50
The Volcanic Carbine 62
New Beginnings For Smith & Wesson: Their Second Partnership 66

chapter five
The New Haven Arms Company's Volcanic Arms 67
General Features of the Firearms 68
The New Haven No. 1 Pocket Pistol 73
The New Haven No. 1 Target Pistol 76
The New Haven No. 2 Navy Pistol 78
The New Haven Carbines 79
The New Haven Pistol-Carbines 83

chapter six
Cased and Boxed Firearms 85
Cased Pistols 86
Cased and Boxed Carbines 92

chapter seven
Special Guns and Special Stories — Presentation and Inscribed Firearms . . . 95
An Inventor's Volcanic Pistol 96
Oliver Winchester's Volcanics 96
A King's Volcanic Pistols 98
An Indian Fighter's Volcanic Pistol 99
A Woman's Volcanic Pistol 100
New Haven Arms Company Carbine Serial Number "0" 102
Benjamin Tyler Henry's New Haven Pistol-Carbine 103

chapter eight
The Volcanic in Service 105
Civil War Volcanics 105
A Lieutenant's Volcanic 105
Major Rockwood's Pistol 108
Major General Butler's Great Southern Expedition 109
The Staten Island "Yellow Fever" War of 1858 111
On To Cuba! 113
A "Small, But Devilish" Volcanic 113

chapter nine
Volcanic Advertising and Promotion 115
Circulars and Broadsides 116
Newspapers 120
City Directories and Almanacs 123

chapter ten
The Henry Rifle and Its Successors — The New Haven Arms Company & Benjamin Tyler Henry 127
The Iron-Frame Henry 131
The Brass-Frame Henry 131
Finally... The Winchester 132

EPILOGUE 135
APPENDIX I: Ammunition & Firearms Patents 136
APPENDIX II: Oliver Winchester's Testimony 142
APPENDIX III: The Venditti Controversy 145
BIBLIOGRAPHY 149
ENDNOTES 151
INDEX 156

preface

The Repeating Dream

The idea of developing a successful repeating weapon has existed since the earliest of times. Centuries of recorded history tell us that armed conflict between individuals, groups, tribes and cultures has prevailed throughout the ages. Thus, it is not unreasonable to assume that some individuals had ideas for improvements that might allow greater and more rapidly applied force for their weapons, thereby increasing the likelihood of success in their battles. The concept of a number of projectiles being launched with the speed of a machine, however, predates firearms, as evidenced by the "polybolos," a repeating ballista invented by Philo of Byzantium in the third century B.C.

Much later, after gunpowder was introduced, cannons predominated for several centuries in the arena of warfare. A repeating magazine firearm was patented in England in 1718 by James Puckle, using multiple loaded chambers each of which had a flintlock attachment for firing. Many other repeating firearms followed, including those with multiple barrels, multiple loaded chambers and methods of repeatedly loading powder and ball from separate magazines built into a repeating flintlock firearm.

As early as 1798, Eli Whitney turned his talents to the manufacture of firearms at Whitneyville, near New Haven, Connecticut. There he invented the cotton gin, which greatly improved the agricultural capability of the South. He achieved another milestone for the North, the principle of standardization or interchangeability of parts, which revolutionized gun making and would give the North a decided advantage in the coming Civil War. Elisha Collier patented a revolving flintlock pistol in 1818, and in 1836, Samuel Colt introduced the first of his famed revolvers, the Paterson pistol. By 1846, Walter Hunt had developed the Volition Repeater, and later, Horace Smith, Daniel Baird Wesson and B. Tyler Henry began the work that would lead to the development of a successful magazine-fed, lever-action, cartridge-repeating firearm.

It is with this background that our story of "The Volcanics" begins...

chapter one

Hunt's Volition Repeater and the Jennings and Smith-Jennings Rifles

WALTER HUNT was a prolific inventor, living in the early and mid-19th century. But, unlike Samuel Colt, Horace Smith, the Wesson brothers and many others, Hunt envisioned and built only one firearm, the "Volition Repeater," along with its intended ammunition, the "Rocket Ball."

He was born the first of thirteen children on July 29, 1796, in Martinsburg, New York. He was educated in a one-room schoolhouse and earned a degree in masonry in 1817. Hunt's early life was spent living as a farmer in the milling community of Lowville in Lewis County, New York. Motivated by the local flax mill's impending closure due to its inability to compete, he invented a mechanical flax spinner that was far more efficient than the previous spinners.

Walter Hunt, 1796–1859

He traveled to New York City in an attempt to raise the capital needed to manufacture the new machine. While there, he saw the opportunity to achieve greater success and moved his family to the city in 1826. By the time his Volition Repeater patent was issued in 1849, he had also invented a foot-operated stagecoach gong, the lock-stitch sewing machine, the fountain pen, a self-closing ink well, nail-cutting machines, rope-making machines, an ice boat and various other useful things. Subsequent to his rifle and ammunition inventions, he went on to invent the safety pin as well as various shirt collar improvements. Hunt is credited with over sixty major inventions during his lifetime.[1] A common theme of many of his ideas centered on devising a way to change a single action into a repetitive one. And so it was with his one attempt at designing a repeating rifle and the ammunition it would fire.

Hunt's role in the development of a repeating firearm began indirectly on January 29, 1846, at a family gathering to celebrate the 19th birthday of the oldest of his four children, George Washington Hunt. After dinner, the conversation drifted to the tense situation along the border between Mexico and Texas and the yet-to-be-declared war. Rumors of the conflicts and atrocities along the border had inflamed citizens to a fever pitch.[2] George had indicated his desire to fulfill his patriotic duty and enlist in the Army, much to his Quaker parents' dismay. At their insistence, George agreed to delay his decision. Perhaps his courting of a young woman further dissuaded him from enlisting. Following an ambush of a reconnoitering party on the Texas side of the Rio Grande on April 25, 1846, Congress acknowledged the reality of the conflict and declared war with Mexico on May 13, 1846.

Despite Hunt's pacifist views and his son's delay in enlistment, he began working on better ammunition and a more efficient repeating firearm to utilize it. Walter Hunt objected to the war, but believed that if the Army had a better gun, the war would end sooner, resulting in fewer deaths.[3] His ideas, as expressed in his firearm and ammunition patents discussed in this chapter, were not only original in the extreme, but were also the beginning of a train of ideas that would eventually lead to the development of a series of breechloading repeating firearms culminating in the Henry and Winchester rifles.[4]

GEORGE ARROWSMITH was a New York City businessman and mechanic who had been associated with Walter Hunt long before 1833 when Hunt debuted his new mechanical sewing machine. He had previously obtained from Hunt the rights to a machine to manufacture pill boxes and a nail manufacturing machine, both of which proved profitable.[5] Arrowsmith wanted half interest in the new sewing machine as payment to manufacture and market it. Hunt needed money immediately for his ongoing living expenses so they ended up trading the sewing machine rights for the rights to the other two previously licensed machines. The swap was made without legal formality.[6] Because of the labor movement's reservations concerning the sewing machine, combined with a cholera epidemic and depression in the mid-1830s, Arrowsmith delayed the patent application for the new lockstitch mechanism. This opened the door for Elias Howe's later claim of inventing it.[7]

HUNT'S VOLITION REPEATER

Arrowsmith and Hunt continued this informal relationship over the years, with Arrowsmith benefitting from Hunt's inventions and Hunt from Arrowsmith's marketing and financial support. When Hunt began his work on ammunition and a firearm in 1846, Arrowsmith was apprised of it early on. Within a few weeks, Hunt revealed to Arrowsmith his Volition Repeater, purported to fire at least twelve shots in succession without reloading. The new ammunition Hunt had developed, his Rocket Ball, was a pre-loaded bullet with a consistent powder load, ensuring better repeatability, shot to shot, compared with the muzzleloaders of the day. Since the Rocket Ball was loaded at the breech end of the repeater's rifled barrel, loading was much faster than a muzzleloaded rifle and far more accurate than the muskets of the day.

They prepared to exhibit the firearm and new ammunition at the American Institute's 19th annual fair in October of 1846 in New York City. However, at the request of the Army, as a patriotic gesture, Hunt agreed not to exhibit his gun until the authorities had a chance to evaluate it. Hunt went on to other projects and allowed Arrowsmith to promote the gun. On September 6, 1846, Arrowsmith submitted the patent application for the Volition Repeater, but he was ill-prepared to promote the firearm.[8] He knew next to nothing about guns, had no means to expedite the patent application, had no political influence and had no contacts with the military purchase board. He did not realize that the Army was too busy with the Mexican War to take immediate interest in a repeating rifle. Arrowsmith knew none of the gun manufacturers and also lacked sufficient capital to promote the manufacture of a rifle.[9]

Despite the fact that the construction of Hunt's rifle was very complicated with its many parts, the operation of the Volition Repeater was not complicated at all.[10] To prepare the gun for firing, it had to be held in a horizontal position with the priming magazine upward so that gravity would position the pill primer correctly. The rear of the two levers, the cocking lever, was employed to move the bolt assembly back as the lever was moved forward, engaging the detent spring, which, when fired, would release the firing pin. With the rear lever still forward, the front lever, or loading lever, was pushed upward with the palm of the other hand, which caused the transfer tube or carrier to drop down to a position in line with the magazine to receive the next Rocket Ball from the magazine. The loaded carrier was then pushed upward by direct hand pressure from below, which brought the cartridge in line with the barrel. At this point, the cocking lever was moved rearward, moving the bolt forward and forcing the cartridge into the chamber. At the full rearward position, the cocking lever

Walter Hunt's Volition Repeater.
Courtesy of the Cody Firearms Museum

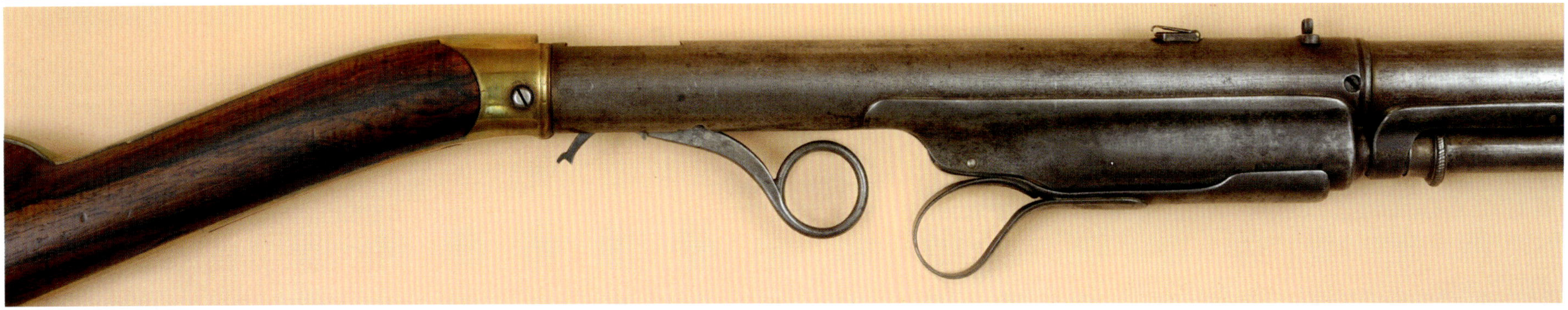

forced a locking lever to move into position to lock the bolt assembly in firing position. A pull of the trigger resulted in the movement of a sear that lifted the detent spring from its retaining pin, releasing the springloaded firing pin which ignited the pill primer. The fire resulting from the primer ignition entered the central opening in the cork base of the Rocket Ball, igniting the powder inside, which forced the bullet through the barrel.[11] The underbarrel magazine of the Hunt rifle was made in two sections. The forward section contained a coil spring in front of a follower, the latter contacting the forward most round in the magazine, which moved the cartridges loaded in the rear section of the magazine to the transfer tube or carrier. This basic design, with subsequent improvements or minor modifications in loading methods, is still used to this day.

At Arrowsmith's urging, some of Hunt's efforts were directed to the selection of the most suitable of his many cartridge designs for use in the repeater with consideration of which was the most practical and economical to manufacture in commercial quantities. Although the tide of the Mexican War turned in favor of the United States, Hunt continued work on his ammunition. He realized that a spherical ball, of the type readily available at numerous shot towers, would not operate in a breechloader.[12] Eventually, toward the end of May 1847, Hunt's Rocket Ball ammunition patent application was ready. It was assigned to Arrowsmith on June 5, 1847, and after Hunt's final review, was filed on June 9, 1847.

Despite the fact that Hunt's gun had been well received, having been awarded a silver medal at the 21st annual fair of the American Institute in 1848, Arrowsmith could not raise capital for its manufacture.[13] He hired Lewis Jennings, a skilled mechanic, to begin production in Arrowsmith's own shop. Jennings and Hunt appear to have worked together in an attempt to build Hunt's original rifle design. Evidence gleaned from the dates of patent applications suggests that, in the fall of 1847, while building the Volition Repeater, Jennings made significant modifications to Hunt's design and that both men ultimately contributed to the first working models of the Jennings repeater.[14]

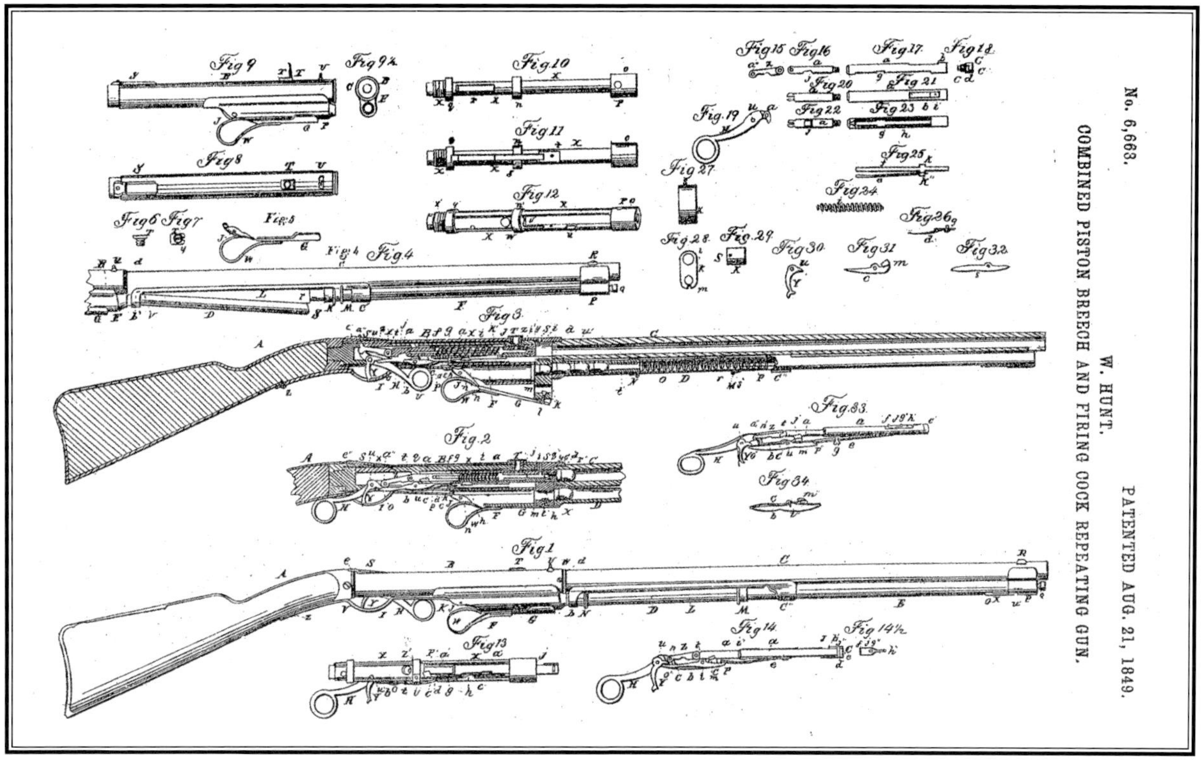

Letters U.S. Patent No. 6,663 for the Volition Repeater. The complexity of the design is readily apparent.

After waiting months for the U.S. patent application process for the Rocket Ball and the Volition Repeater to move forward, Arrowsmith, out of frustration, applied for a patent in England. This patent application was for a combination of Hunt's Rocket Ball cartridge and the Jennings improvements to the repeating rifle.[15] English Patent No. 11,994 was granted on December 10, 1847, and Hunt subsequently assigned the English rights to Stephen Taylor, located in London. In a futile attempt to expedite action on the U.S. applications,[16] and with the English patent in hand, Arrowsmith hired Charles M. Keller, a Washington patent attorney.

R. Bruce McDowell, author of *Evolution of the Winchester*, theorizes that, in addition to the greater cost to produce, the original Hunt Volition Repeater was never manufactured in significant numbers because it was undergoing redesign by Lewis Jennings long before the Hunt U.S. patent was granted.[17] It is estimated that only a few Hunt rifles, and perhaps only the U.S. patent model, were ever built. The application for the Volition Repeater was delayed, in part, because of several interference claims by other parties. These issues were eventually resolved, and U.S. Patent No. 6,663 was granted on August 21, 1849.[18] Today, the only known Hunt Volition Repeater resides in the Cody Firearms Museum in Cody, Wyoming.

During the almost three years that the Volition Repeater patent application was pending at the U.S. Patent Office, Hunt continued to work on improvements to his cartridges. On December 21, 1847, he filed an addendum covering a new type of cartridge, superior to the Rocket Ball, that he had assigned to Arrowsmith. Repeated tests demonstrated the superiority of the cartridge improvements and Arrowsmith demanded assignment of this patent to him, for no additional payment. When Hunt refused, Arrowsmith, through his attorney, Charles Keller, notified the U.S. Patent Office that Hunt's application was merely a variation of form of the previously applied-for Rocket Ball patent. As Hunt had not yet filed an application for a patent on this improvement, the U.S. Patent Office could take no action. Hunt waited until July of 1848 to file his application for an "Improved Method of Attaching a Ball to a Wooden Cartridge."[19]

This dispute would end the relationship between Arrowsmith and Hunt. With three of Hunt's applications on file, Arrowsmith asserted that the improved wooden cartridge was only an improvement on the previously assigned metallic cartridge, the Rocket Ball. The Patent Commissioner subsequently decided that both applications were novel and original and qualified for patents. Accordingly, he issued U.S. Patent No. 5,699 to Walter Hunt and U.S. Patent No. 5,701 on assignment to Arrowsmith on August 10, 1848.

PATENT TIMELINE

Patent	Description	Date of Application	Date of Issue
U.S. 5,701	Hunt Rocket Ball	June 9, 1847	Aug. 10, 1848
U.S. 6,663	Hunt Repeating Rifle	Sept. 6, 1847	Aug. 21, 1849
Eng. 11,994	Combination of Hunt Rocket Ball and Jennings Repeating Rifle	Unknown	Dec. 10, 1847
U.S. 5,699	Hunt's Cartridge Improvement	July 1848	Aug. 10, 1848

Because the U.S. Patent No. 5,699 patent text does not relate what Hunt really had in mind for this cartridge design, we will probably never substantiate his intentions. It did, however, protect a definite centerfire-type cartridge consisting of a case with a central base opening containing "gun cotton," the powder charge and a projectile. Gun cotton was a recent development made by treating cotton with equal parts of nitric and sulfuric acids and then slowly drying the cotton wool. Schönbein, a German-Swiss chemist, had demonstrated it in 1846 as an explosive propellant for small arms.[20] Hunt may have intended this as the primer. It cannot be established that this cartridge was ever fully recognized or produced.[21] For reasons unknown, Hunt later assigned this patent to William R. Palmer, Courtlandt Palmer's brother, on February 16, 1850. This same wooden cartridge patent is later listed among the patents consigned to Oliver Winchester, under which the New Haven Arms Company held the rights to manufacture arms when he established the company in April 1857.[22]

Despite the drawbacks of the Volition Repeater, Hunt's design of a lever-operated rifle with an underbarrel magazine, holding fixed ammunition cartridges was a significant beginning. In addition, his design of what he termed a "transferring tube," subsequently called the "carrier" by Jennings, to elevate the cartridge from the magazine to a position where the bolt could move the round into the chamber, was revolutionary and unique. Hunt's ammunition experiments led him to design the Rocket Ball, a melding of propellant as part of the projectile, and while it had its drawbacks, it was a significant step on the path to a complete cartridge. Hunt's other ammunition patent, for a wooden-cased cartridge, is very intriguing. One could speculate that, if this concept had been pursued at the time, centerfire cartridge ammunition may have been developed years ahead of when it was actually produced. Impetus for the rifle and ammunition development passed with the end of the Mexican war, and Hunt moved on to devote his time to other inventions. On June 8, 1859, fewer than eight weeks short of his 64th birthday, Walter Hunt died of pneumonia while in his workshop. On June 13, 1859, the *New York Tribune* wrote of Hunt's death:

"For more than forty years, he has been known as an experiment in the arts. Whether in mechanical, chemistry, electricity or metallic composition, he was always at home: and, probably in all, he tried more experiments than any other inventor."[23]

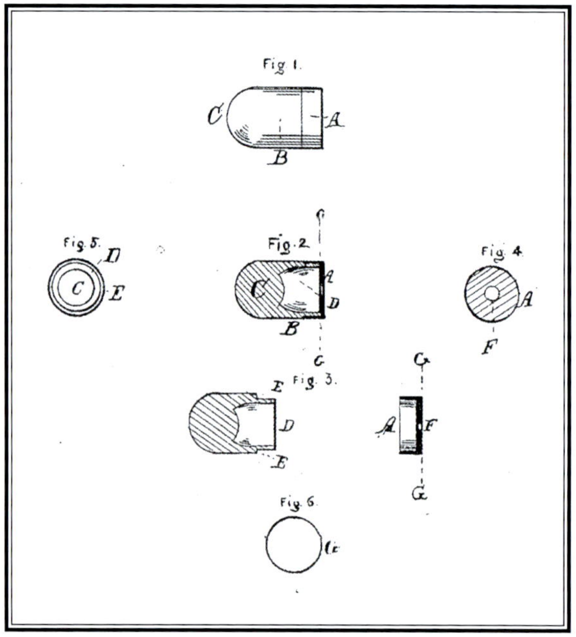

Hunt's Letters U.S. Patent No. 5,701 for a "Cartridge," which was assigned to George Arrowsmith prior to issuance.

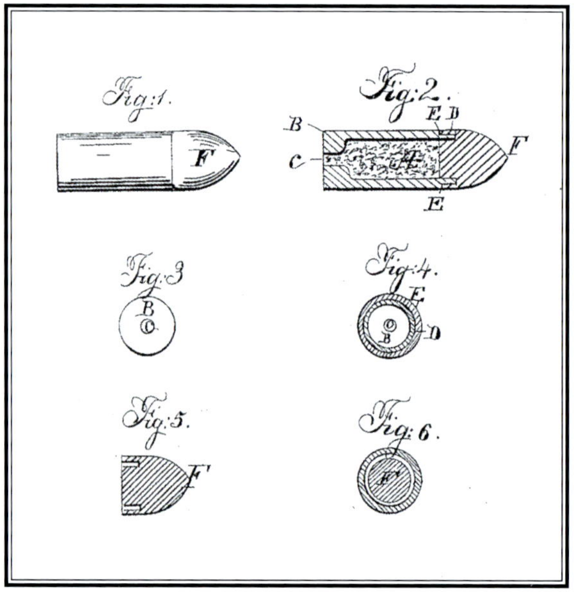

Hunt's Letters U.S. Patent 5,699 for an "Improved Method of Attaching a Ball to a Wooden Cartridge."

THE JENNINGS RIFLE

At some point in early or mid-1847, Lewis Jennings, an inventor and expert model maker employed by George Arrowsmith, joined with Walter Hunt to build and perfect the Volition Repeater in Arrowsmith's shop in New York City. Jennings' improvements substituted a standard side hammerlock to ignite the pill primer rather than the bolt-enclosed spring-loaded firing pin action of Hunt's design. As the hammer was drawn back to half cock, it loosened the breech pin from the barrel by breaking the seal remaining from the previous Rocket Ball ignition. Jennings also simplified the Hunt design, which consisted of a series of levers operated by the cocking lever with a single-ring trigger attached to a rack-and-pinion system. The ring trigger, upon moving forward, simultaneously

First Model Jennings rifle, serial number 341, with original pill primer ignition. Stephen Rutter collection

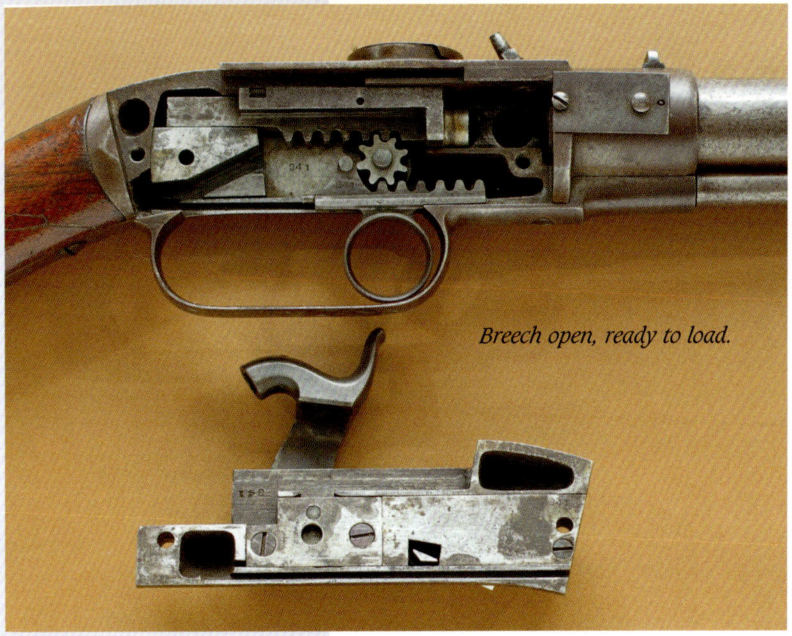

Breech open, ready to load.

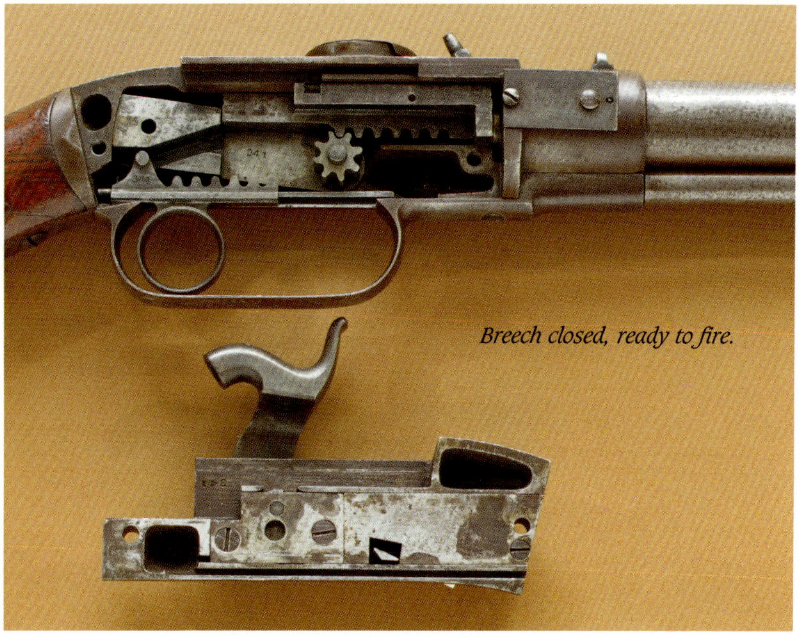

Breech closed, ready to fire.

Serial number 341 with sideplate removed. Stephen Rutter collection

moved the breech pin rearward, thereby opening the breech completely. The trigger bar also indirectly moved the carrier elevating lever, thereby elevating a cartridge from the underbarrel magazine to be in line with the chamber. The rear movement of the same ring trigger caused the bolt to push the cartridge from

Left: *The left side of the frame is inscribed "PATENT 1849/ C.P. DIXON AGENT/NEW YORK".* Stephen Rutter collection

Below, top: *First Model Jennings rifle, serial number 20. Note the unusual brass patch box.* Stephen Rutter collection

Below, bottom: *Third Model Jennings rifle, serial number 02, a factory-produced muzzleloader with a non-factory adjustable rear sight. Percussion cap ignition.* Stephen Rutter collection

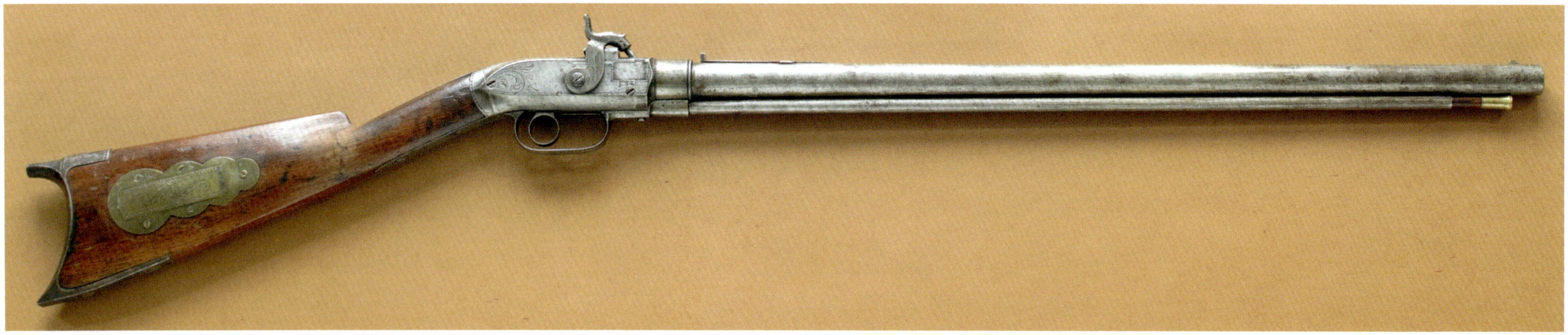

the carrier into the chamber and then lower the carrier into position to receive the next Rocket Ball from the magazine. The initial rearward movement of the breech pin caused a single pill primer to fall from the primer magazine, located on top of the receiver, into position to be ignited by the released hammer. Returning the ring to the rear position locked the breech and, after drawing the hammer to full cock, fired the charge.

The only Hunt design elements that were retained in the Jennings improvements were the sliding internal bolt or breech pin, the tubular cartridge magazine, which was under and in line with the barrel and the percussion pill magazine and the feed system. While similar to the Volition Repeater, the bolt locking lever in the Jennings rifle was much stronger. The carrier, which transferred the cartridge from the magazine to the chamber of the barrel, was made action operated, though it could be overridden by a blocking pin to prevent the loading of another cartridge, in the event of a misfire that would require re-cocking the hammer.

Unlike the coil spring and follower design of the Hunt cartridge magazine, the magazine designed by Jennings was an extremely complex double-ratchet system. It was delicate and undependable, due to its reliance upon gravity rather than a positive mechanical action.[24] Jennings applied for a U.S. patent on August 2, 1848, and it was issued on December 25, 1849. Jennings assigned his patent to George Arrowsmith.[25] Arrowsmith recognized that he had neither the skill nor the capital needed to properly finance the manufacture of guns. He subsequently sold his rifle and ammunition patent interests to Courtlandt Palmer, a leading New York City hardware merchant, industrialist and promoter, for the sum of $10,000.[26]

Courtlandt Palmer, 1800–1874
Courtesy of the Stonington Historical Portrait Collection, Stonington Historical Society, Stonington, Conn.

COURTLANDT PALMER was born in Stonington, Connecticut, on November 11, 1800. After a common school education, he began a hardware business in New York City in which he was very successful. The financial crisis of 1837, however, was disastrous for his firm. He invested his remaining capital in real estate, which proved to be his financial salvation. For several years he was president of the Providence & Stonington Railroad and was one of the founders of the Safe Deposit Company of New York. Over his business lifetime he was a director of several businesses, and at the time of his death in 1874, he was included among the richest men of New York.

After acquiring Hunt's two patents as well as the patent for the Jennings rifle, Courtlandt Palmer contacted Richard S. Lawrence, a principle in Robbins & Lawrence Company, of Windsor, Vermont, the largest non-government firearms manufacturer in the country at the time. It was one of the first factories to manufacture interchangeable parts for firearms and to produce guns under contract for a number of other companies. From 1842 until 1856, the superintendent at Robbins & Lawrence was Benjamin Tyler Henry,[27] a master machinist, born in Claremont, New Hampshire, on March 22, 1821.[28] Palmer entered into a contract with Robbins & Lawrence for the production of 5,000 Jennings rifles, though most authorities find evidence that fewer than 1,000 Jennings rifles were ever actually manufactured.[29] Jennings rifle production was begun early in 1850, and it soon became apparent to Palmer that further improvements in design were needed for the rifle to be a successful repeater.[30] During development of the Jennings rifle, it was discovered that, if the Rocket Ball had grooves cut into the outside surface that were filled with tallow, there would be reduced leading of the barrel and improvement in accuracy.[31] Despite this important advancement of bullet lubrication, which resulted in greater accuracy, C.P. Dixon of New York, whom Palmer had hired as an agent, experienced difficulty with promotion and sales. He managed to have numerous endorsements published in various newspapers and magazines of the day, but failed to impress the U.S. Ordnance Department during their trials in October 1850.[32] The Department's conclusion on the Jennings rifle was: "The Board are of the opinion that the principle of construction of these arms is such as would not be suitable for military service."[33]

Very few Jennings repeating rifles were ever built, and most likely only for testing and evaluation.[34, 35] It appears that most of the initial production was of single-shot breechloaders and, late in 1851, when it became evident that the Jennings rifle had gained little popularity, the remaining parts were assembled as muzzleloading rifles.[36] The majority of single-shot breechloaders produced by Robbins & Lawrence were later converted to percussion muzzleloaders by owners, gunsmiths, mechanics or blacksmiths; no two are alike.[37] The Jennings single-shot breechloaders and factory muzzleloaders usually had cleaning or ramrod tubes under the barrel. The rods were made of hardwood with a brass rammer cap on the forward end and a brass patch tip on the end stored in the bolster of the triggerguard. A unique feature of the Jennings cleaning rod is that the patch tip has two areas of threads to allow it to be extended to reach the entire length of the barrel and then retracted, to fit beneath the barrel.[38] Three models of the Jennings rifle are generally accepted as having been produced between 1850 and 1852.[39]

1st model:	Single-shot breechloader with an underbarrel cleaning rod tube. 1850–1853
2nd model:	Repeating rifle, tubular magazine, attached with screws beneath the barrel, loaded from the front. Very few were made. 1851–1853
3rd model:	Factory muzzleloader and any non-factory conversions to muzzleloaders. 1852–1853

In both the single-shot breechloading and muzzleloading Jennings rifles, numerous variations have been noted. These include the receiver shape, pill primer magazine shape as well as the style and operation of the loading port covers. R. Bruce McDowell has compiled a number of these variations after examining a total of twenty-seven Jennings breechloaders.[40]

VARIATIONS IN JENNINGS BREECHLOADERS

(as found in 27 Jennings rifles examined) from McDowell, *The Evolution of the Winchester*

Large-frame	2	Trigger bar with wide cog (as in the repeater)	0	Stock inlay on left side	16
Small-frame	25	Stock inlay on right side	10	Flat hammer	12
Full-length trigger bar	12	Carrier retained and operative	1	Rounded hammer	15
Shortened trigger bar	15	Round cover on pill magazine	22	Blued barrels	18
Bolt locking lever with front pivot	19	Square (hinged) cover on pill magazine	2	Browned barrels	9
Blued lock frames	21	Bolt-locking lever with rear pivot	8	Open-style sporting sights	23
Browned lock frames	6	Screw-elevated "V" rear sight	4	Blued buttplates	14
Pill magazine retained	24	Standard scroll engraving	27	Browned buttplates	9
Pill magazine rotating pan intact	4	Full standard markings	26	Case-hardened buttplates	4
Ring trigger spring added	11	Single-digit serial number	1	Screw-swivel loading port cover	19
First-style buttplate	21	Double-digit serial number	15	Sliding loading port cover (beveled tracks)	7
Third-style buttplate	6	Triple-digit serial number	11	German silver oval stock inlay	26

Jennings' serial numbers started with number "1" and ran consecutively through 1st and 2nd model variations and modifications until the Smith-Jennings repeaters (discussed later) and the factory Jennings muzzleloaders were manufactured. Both of these began with a new series of serial numbers.[41] Because the Jennings rifle proved unsatisfactory as a repeating rifle, Palmer hired Horace Smith to improve the Jennings rifle design and to supervise its production at the Robbins & Lawrence Company.

The path to the Winchester rifle was not direct and not without its twists and turns. Just as Hunt's Volition Repeater began a collection of patents, including the Jennings patent, which later came to be owned by Oliver Winchester, certain features of the Jennings rifle such as the sliding bolt, the underbarrel magazine and the cartridge carrier, were eventually incorporated into the Winchester rifle.

Noted historian and author Herbert Houze has stated that: "Once touted as the greatest contribution to the development of firearms, the Jennings Patent rifle quickly slipped into oblivion and is remembered today, somewhat incorrectly, as an unsuccessful precursor to the lever action Winchester rifle." How did this come about? Houze further tells us that, in 1922, in a paper presented to the Vermont Historical Society, Guy Hubbard claimed, "Mr. Henry took this early Windsor rifle, [the Jennings] to New Haven and made a few changes...he made hundreds of them during the Civil War as the Henry Rifle. In 1866, he and Gov. Winchester founded the Winchester Repeating Arms Company, and the Jennings Rifle became the far famed Winchester Repeater..." Houze concludes, "thus was born the story that the Jennings was the direct antecedent of the later Winchester."[42] That said, had it not been for the problems encountered in producing the Jennings rifle, Horace Smith may never have been involved in the process.

HORACE SMITH was born in Cheshire, Massachusetts, on October 28, 1808. At the age of four, his family moved to Springfield where his father had accepted a job at the Water Shop, part of the Springfield Armory complex. In 1824, at the age of 16, Horace joined his father at the Armory as an apprentice to the bayonet forger. After four years in that position, he became a journeyman tool and die maker.[43] This training provided the foundation for his future success. It is said that he worked at the Armory for 18 years and then, in 1842, at the age of 34, he moved to Newton, Connecticut. He then worked for 15 months making tools for Eli Whitney in New Haven, Connecticut. After that, he worked at Allen & Thurber in Norwich, Connecticut, until 1848, when he opened his own shop in Norwich.[44] He was perhaps the most respected gunsmith in New England at the time when, in 1850, he was hired by Courtlandt Palmer to join Jennings at Robbins & Lawrence to further improve the Jennings rifle design.

Also in 1850, Daniel B. Wesson was employed as superintendent at the Leonard Pistol Works where he was assigned to Robbins & Lawrence in Windsor, Vermont, to supervise the manufacture of a pepperbox pistol. It was at Robbins & Lawrence where Benjamin Tyler Henry was first introduced to the Jennings breechloading repeating rifle. Thus the stage was set for the ongoing collaboration of Jennings, Smith, Wesson and Henry in their mutual quest for a repeating firearm.

THE SMITH-JENNINGS REPEATER

Smith made significant improvements to the Jennings design, changing from the rack-and-pinion system of the Jennings rifle to a single lever that operated the breech pin directly by moving in a recess on the side of the breech pin. The single lever also directly contacted the lever that elevated and lowered the carrier to deliver a new cartridge to the breech.[45] He also simplified the magazine system, eliminating the complicated ratchet design of Jennings. Smith's underbarrel magazine used a front-loaded coil

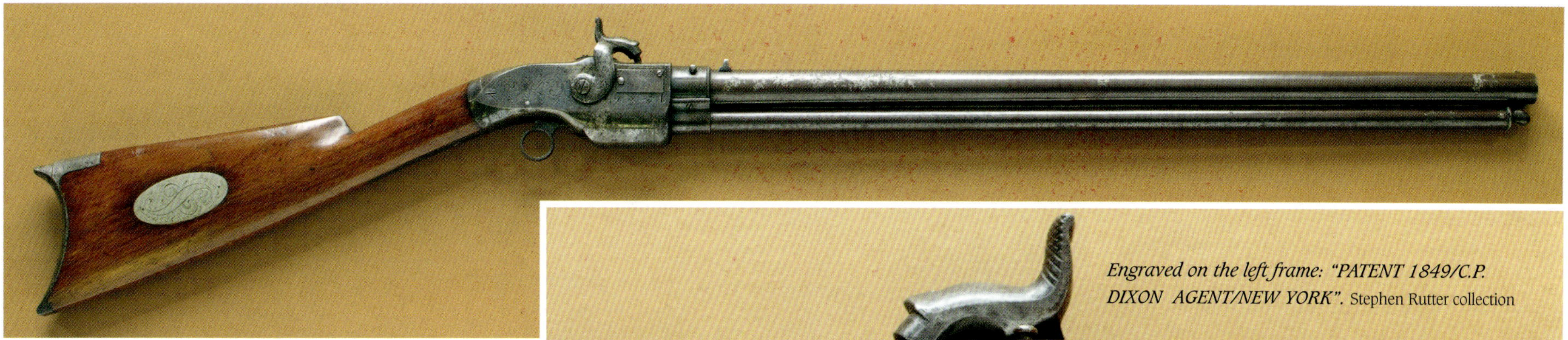

The Smith-Jennings Second Model rifle, serial number 3C.

Engraved on the left frame: "PATENT 1849/C.P. DIXON AGENT/NEW YORK". Stephen Rutter collection

Horace Smith, 1808–1893

spring and follower system accessed by simply opening the muzzle end of the magazine tube. Smith received U.S. Patent No. 8,317 on August 26, 1851, and assigned it to Palmer. A number of variations and modifications have been noted in the Smith-Jennings rifle, and it has generally been agreed to classify them into three models. All were capable of use as a repeating rifle as well as a single shot by simply loading the ammunition through the receiver loading port. They apparently proved to be successful repeating rifles, as none converted to muzzleloaders have been identified, as was the case with the Jennings rifle.[46]

Since Smith's patent was an improvement of the Jennings rifle, it has become known today as the Smith-Jennings rifle, though a number of contemporary articles refer to it as the "Jennings' Patent Rifle."[47] Like the Hunt and Jennings repeaters, the Smith-Jennings repeating rifle was designed to shoot the Hunt Rocket Ball, a .54 caliber conical lead bullet, loaded from the rear with black powder and sealed by a cork disk with a small hole in its center to allow a separate pill primer to ignite the charge. By 1852, it became apparent that even the Smith-Jennings repeating rifle systems were too complicated and not powerful enough to become commercially successful. Production was abandoned after the remaining parts were assembled into completed rifles.

The Robbins & Lawrence Armory in Windsor, Vermont, built in 1846.

THE SMITH-JENNINGS EXPERIMENTAL PISTOL

Shown is an interesting and unique Smith-Jennings pill lock repeating pistol in .36 caliber with an 8½-inch barrel and tubular magazine. Its operation is identical to the repeating rifle. The bag grips portend the future grips of the later small-frame Smith & Wesson model pistol. This experimental pistol has features which indicate that it was produced at the time of the First Model Smith-Jennings rifle.[48]

PRODUCTION NUMBERS

Hunt Repeating Rifle: One known, located in the Winchester Museum, Cody, Wyoming.[49]

Jennings Rifle: It has been estimated that fewer than 1,000 rifles in all three variations were produced.[50]

Serial numbers began with no. 1 and ran through the 2nd model. They started over with no. 1 for the factory muzzleloaders.

Smith-Jennings Rifle: Estimates are that only about 1,000 rifles, in all three models, were produced.[51]

Serial numbers began with no. 1 for each of the three models.

While the Hunt, Jennings and Smith-Jennings rifle designs all contributed to the development of the tubular magazine Winchester rifle, their greatest contribution lies in the fact that, through these firearms, Horace Smith, Daniel Wesson and Benjamin Tyler Henry all began work on a lever-action repeater and its suitable ammunition.[52]

Gerald Klaz collection

chapter two

The Smith & Wesson Lever Action Firearms

THE FIRST "VOLCANICS"

The lack of a suitable cartridge for the Smith-Jennings rifle contributed significantly to its poor sales and the end of its production. What was needed and unavailable at the time was a firearm that would function as a repeater that could fire a self-contained cartridge. Upon completion of his contract with Courtlandt Palmer in 1852, Horace Smith appears to have been employed by Allen, Brown & Luther of Worcester, Massachusetts. Daniel B. Wesson was also an employee of that firm at the same time, so it is plausible that their contact there fostered continued collaboration on a repeating firearm.[1]

It has been speculated that both men concluded that the failure of prior attempts at designing such a firearm was primarily due to the inadequacies of the Rocket Ball ammunition. Their efforts were directed at solving that problem during their informal partnership, which lasted from 1852 until 1854.[2]

Daniel Baird Wesson, 1825–1906

DANIEL BAIRD WESSON was born in Worcester, Massachusetts, on May 18, 1825, to Rufus and Betsy Baird Wesson. The family was of limited means as his father worked the land and manufactured wooden plows. The young Wesson helped his father on the farm when he was not attending public school. His older brother, Edwin, was a gunsmith and made rifles in his gun shop in Grafton, Massachusetts, where he gained a reputation as a manufacturer of fine and very accurate target rifles. By the age of 17, Daniel's interest in guns and their development led him to an apprenticeship with Edwin to learn the trade of gun making. The indenture agreement was signed by Edwin and his father on December 12, 1842. The terms of the agreement were that Edwin would pay his father the sum of $250 within the next three and one-half years and that Edwin agreed to supply room and board. Daniel was granted three months for schooling and was to be provided the necessary instruction to make him a master gunsmith by the time he was 21 years old.[3] For his part, Daniel agreed to grant Edwin his time and service and to apply himself industriously in his brother's gun shop in Northboro, Massachusetts. Daniel thus began his career as a designer and manufacturer.

While Edwin traveled, Daniel was left in charge, and during his indenture, he met important individuals in the firearms business, including Samuel Colt and Samuel H. Walker, a captain of the U.S. Mounted Rifles during the Mexican-American War. The latter had ordered a rifle from the Wessons, and correspondence in February of 1847 indicated that Walker wished to order 1,000 rifles for his regiment. The order, however, was never actually completed. On November 22, 1848, Edwin and Daniel signed an agreement acknowledging completion of his apprenticeship. The great promise of the brothers' enterprise came to an abrupt end on January 29, 1849, when Edwin unexpectedly died of a heart attack.[4] Financial difficulties prevailed, and on November 22, 1849, the Wesson rifle business was publicly auctioned and acquired by the Massachusetts Arms Company.

Assisted by his other brothers Franklin and Martin, Daniel Wesson continued to build percussion rifles and pistols in Hartford, Connecticut, until October 1850. He was employed by several companies after he moved back to Worcester. He worked for a short time at the Leonard Pistol Works as factory superintendent and after that was employed by the firm of Allen, Brown and Luther, manufacturer of machine tools and gun barrels. Here he became familiar with the style of Allen & Thurber pepperbox pistols. Prior to the beginning of the Smith & Wesson informal partnership in 1852, he also worked for a man from Tennessee named Porter to improve his newly patented turret-style revolving rifle.[5]

By 1852, Walter Hunt, Lewis Jennings, George Arrowsmith and Richard Lawrence disappear from the scene, but Courtlandt Palmer, Horace Smith, Daniel B. Wesson and B. Tyler Henry continued to work on repeating firearms. There are conflicting opinions as to how and where Smith and Wesson first met. Some claim they met while both men worked at the Windsor factory of Robbins & Lawrence between 1850 and 1851 during the production of the Jennings rifle. Others say they first met at the Worcester rifle factory of Allen, Brown and Luther. They may have met even earlier than 1850, since both were stockholders of the Massachusetts Arms Company.[6]

While working at Robbins & Lawrence, Daniel Wesson and Horace Smith, both knowledgeable gun makers, were aware of improvements in firearms around the world. One such improvement was a self-contained cartridge patented in France in 1846 by Louis Auguste Flobert,[7] which was simply a round ball in the end of a percussion cap. After purchasing a quantity of percussion caps and buckshot from Crittenden & Tibbals, Wesson began experimenting with it in single-shot pistols during 1851 and 1852.[8] Conjecture by McDowell in his 1985 book repeated the undocumented story with, "It is generally believed that Horace Smith was sent to London by Courtlandt Palmer to study developments of the European gun makers."[9] This idea was propagated in an 1958 article by Graham Burnside, repeated in a book published in 1962 by William Edwards in which Smith was reported to have traveled to France and then to have been an exhibitor at the London Great Exhibition of 1851. In a subsequent article published in 1973, however, Frank Sellers refutes these claims with documentation that neither Smith nor Wesson had left this country in 1851.[10] It is this alleged exposure to European arms ideas that seems to have led some to conjecture that the Venditti repeating lever-action pistol preceded Smith and Wesson's pistol. The actual evidence, however, supports the assertion that the Venditti pistol was a copy of the New Haven Arms Company Volcanic pistol. (See Appendix III)

On May 10, 1853, Smith and Wesson applied for a patent entitled "IMPROVEMENT IN CARTRIDGES" that added a chamber within the Flobert-style cartridge, directly behind the bullet, to contain tallow to improve the cartridge's function. Such a cartridge was also de-

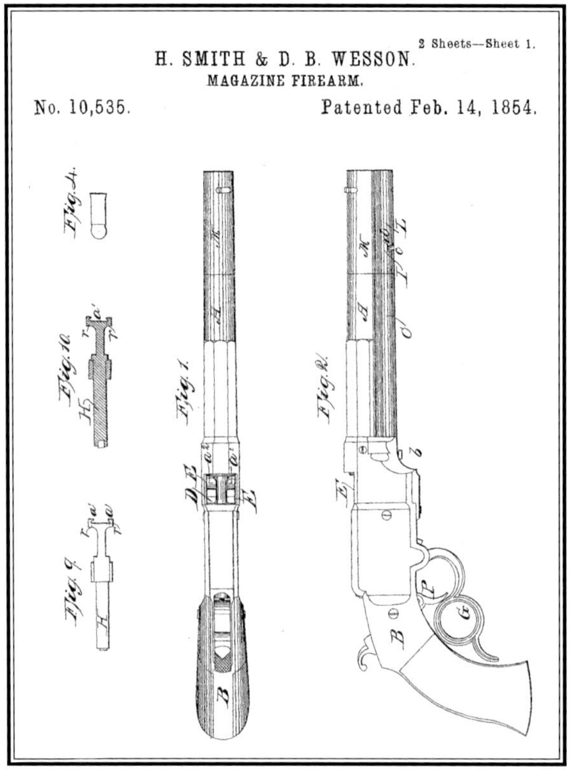

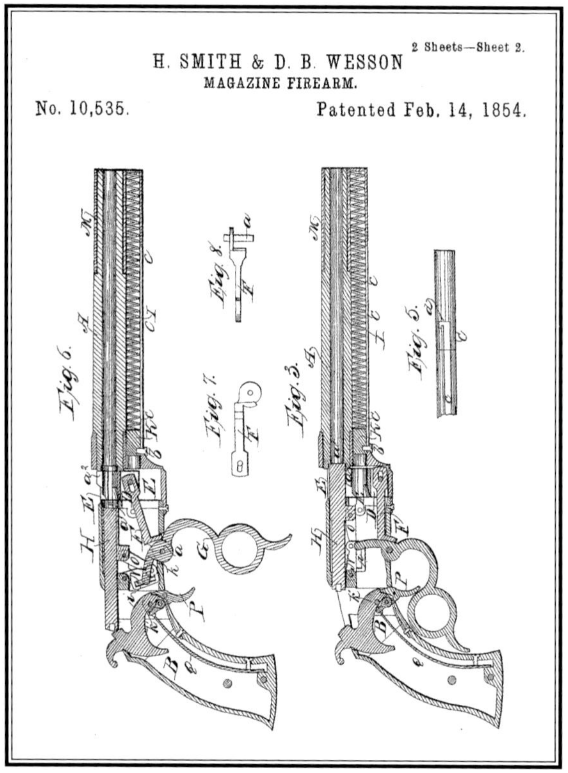

Smith and Wesson U.S. Patent No. 10,535 for a "MAGAZINE FIREARM", issued February 14, 1854.
From McDowell, *Evolution of the Winchester.*

Smith & Wesson U.S. Patent 10,535 issued February 14, 1854. Arrow points to the Flobert-style cartridge.
From Jinks, *History of Smith & Wesson*

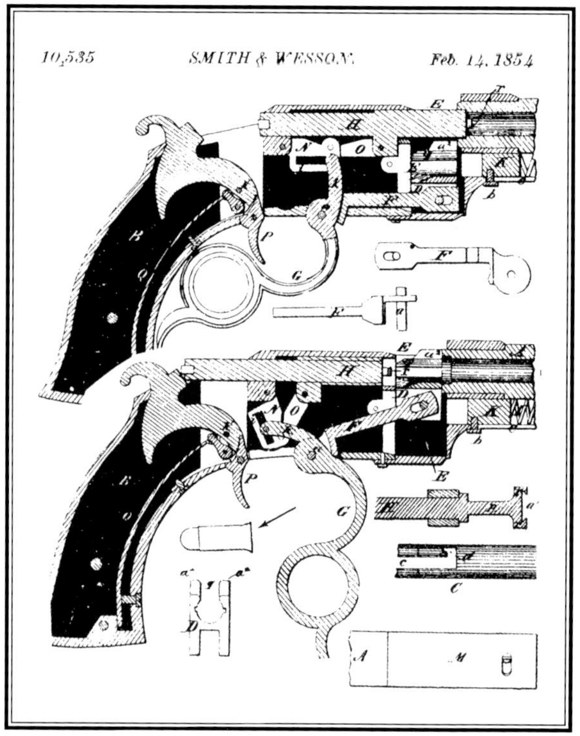

scribed and illustrated by Smith and Wesson in their U.S. Patent application filed just two weeks later, on May 25, 1853, which depicted a "MAGAZINE FIREARM". Both patents were eventually granted, though in reverse order of their application. Flobert's self-contained cartridge patents of 1846 and 1849 became an issue during the application process since the cartridges were similar. After an extensive examination period, U.S. Patent No. 11,496 for an improved cartridge was finally issued to the partners on August 8, 1854.[11]

Their U.S. Patent No. 10,535 of February 14, 1854, "IMPROVEMENT IN FIREARMS", states:

"Fig. 4 is a side view of the kind of cartridge to be used therein, it being the same as used in that species of pistol usually termed the "SALOON PISTOL," it being, as we believe a French invention. The cartridge has its case of thin copper, and with a projecting bottom that contains the percussion powder, the ball being inserted in the other end or top of the case."

The patent illustrated a new firearm that was of a toggle-link design, actuated by a finger lever that cocked the hammer and trigger by the bolt as the finger lever was moved forward, and then loaded a cartridge into the chamber as the lever returned rearward. The front face of the bolt had a dovetail extractor to fit the pictured cartridge. It is unknown whether the mechanism was invented by Horace Smith and Daniel Wesson alone or whether B. Tyler Henry played a role in its design.

BENJAMIN TYLER HENRY was born in Claremont, New Hampshire, on March 22, 1821. In 1837, at the age of 16, he began an apprenticeship at the custom gun shop of J.B. Ripley & Company in Claremont. From there he moved to Windsor, Vermont, where he worked as a machinist at a gun shop. Henry was subsequently employed at the U.S. National Armory in Springfield, Massachusetts, and then at Robbins & Lawrence in Windsor, Vermont. He was the shop foreman at Robbins & Lawrence beginning in 1842 and continuing through the period that Hunt, Jennings, Smith and Wesson worked there in 1850 and 1851. Some sources maintain that he remained there until the reorganization of Robbins & Lawrence in 1856.[12] Other sources indicate that the Smith & Wesson Company hired him in 1854 as an employee and that he returned to Robbins & Lawrence in June of 1855 upon the sale of the Smith & Wesson Company to the Volcanic Repeating Arms Company. When the Volcanic Repeating Arms Company was established, Daniel Wesson is said to have remained until February 1856 as plant manager. He was followed by William Hicks.[13] Some authors state that Winchester hired Henry as plant manager for the New Haven Arms Company starting on May 1, 1857,[14] and others state that Henry replaced Hollis Smith as superintendent there in April or May of 1858.[15] Henry's involvement in the toggle-link design of the lever pistol, patented by Smith & Wesson in 1854, ranges from probable collaboration with Smith & Wesson while they were at Robbins & Lawrence to having actually developed the idea himself.

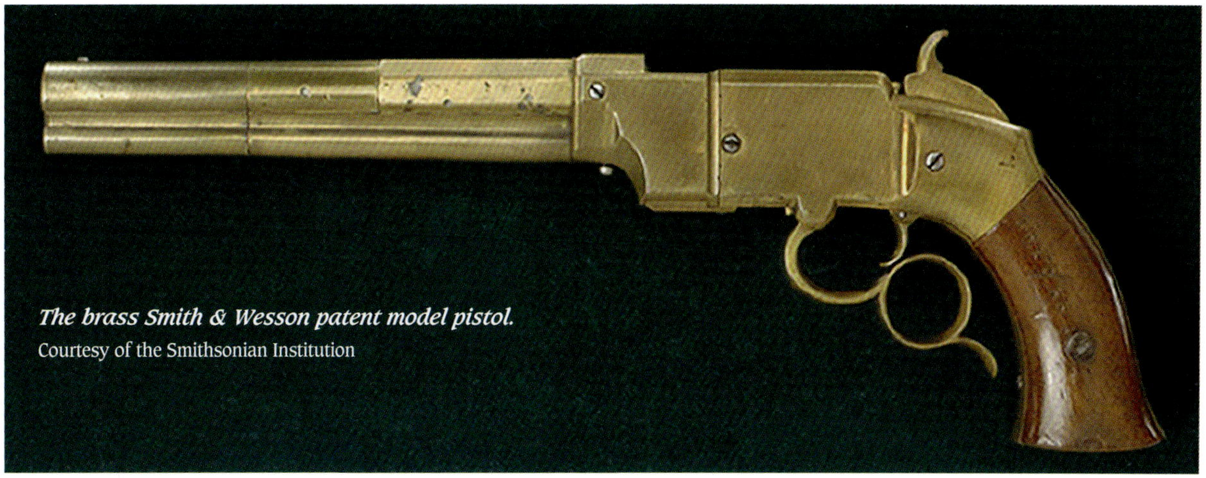

The brass Smith & Wesson patent model pistol.
Courtesy of the Smithsonian Institution

Benjamin Tyler Henry 1821–1898

In his publication, *Famous Guns from the Winchester Collection,* Hank Weiand Bowman states that, during the time in which Smith was working for Palmer to improve the Jennings rifle design, Henry approached Smith, in confidence, and told him about his improved system design in an attempt to obtain the necessary financing for a patent application. This Henry improvement was an action-lever that pivoted at the front and, when pulled back, would force a fresh cartridge into the chamber of the barrel and cock the action. It is possible that Smith & Wesson's 1854 patent reflected some of Henry's design ideas,[16] but Col. B.R. Lewis, writing in 1957, goes a step further and states that the toggle-link idea was originally Henry's, and that it was sold to and later patented by Smith & Wesson.[17] If Smith & Wesson had purchased the idea from Henry, one wonders why the delay from 1851 to May 24, 1853, when the Smith & Wesson U.S. Patent No. 10,535 was filed? Perhaps the seed of an idea had been planted in Smith's mind by Henry, and along with Wesson, they developed the final design that was eventually patented. In fact, the toggle-link design of the patent model of the Smith & Wesson Lever Pistol differs in how the two links join from those that were then manufactured and sold, indicating continued refinement of the design by Smith & Wesson.[18]

The stage was set for the manufacture of the new firearm but the question of how to finance it remained. After working on the new design for almost two years, they had incurred large debts and had only the two patents to show for their endeavors. Courtland Palmer was well aware of the shortcomings of the Jennings and Smith-Jennings rifles and wished to recoup his prior investments in those failed firearms. In addition, the proposed gun had features which came close to infringing on Palmer's own patents, and to protect his interests, he decided to join Smith and Wesson's new partnership. He was particularly interested in the promising design of the new firearm after U.S. Patent No. 10,535 had been issued on February 14, 1854.

Horace Smith, Daniel Wesson and Courtlandt Palmer agreed to a formal partnership, the Smith & Wesson Company, which began on June 20, 1854. Smith brought to the partnership his gun shop in Norwich, Connecticut, along with his share of the patents jointly owned with Wesson for the new self-contained cartridge and the new toggle-link design. Wesson contributed his share of the two patents and firearm parts produced in 1853, estimated to be enough for the production of between 200 and 250 large-frame pistols.[19] For his part, Palmer provided $10,000 in capital and assigned his Hunt, Jennings and Smith-Jennings patents to the new company. The company was short-lived, but in the year of its existence, it produced approximately 1,700 pistols of which fewer than 500 were the large-frame .41 caliber model and about 1,200 were the small-frame .31 caliber models.[20] A few lever action rifles in .50 caliber were also produced. Smith & Wesson's self-contained cartridge was never successfully used in the new firearms. An improved version of the Hunt Rocket Ball was used instead.

ORIGIN OF THE NAME "VOLCANIC"

"Volcanic" was not the name given by Horace Smith or Daniel Wesson to their pistols, and it was never used by the Smith & Wesson Company. Roy Jinks, in his book, *History of Smith & Wesson*, gives this origin of the name:

"Because of its ability to be fired rapidly, the gun was nicknamed the "Volcanic" since it compared to the fiery eruption of a volcano. It is popular belief that the name was given by a magazine entitled Scientific American *when it reviewed the gun in 1854."*[21]

The Volcanic Repeating Arms Company was incorporated in July of 1855, and the new company was the first to market and sell their guns as "Volcanics." Herbert Houze has said that the earliest use of the name was a press piece that was written at the manufacturer's request and printed in several newspapers including a March 1855 issue of the *New Haven Palladium.* The article, entitled "Volcanic Repeating Pistol" then appeared in the *Republican Compiler* of Gettysburg, Pennsylvania, on March 26, 1855, and was shortly thereafter published in the Platteville, Wisconsin, *Independent American*, April 2, 1855:

"Volcanic Repeating Pistol"

"We have seen and fired a pistol recently invented and patented which bids fair to excel everything in that line that has yet been offered to the public attention. It seems to combine all that could be desired in such a weapon. Colt's pistol, compared with it, seems like a distortion; or a clumsy, uncouth, and ridiculous affair for a fire arm. The Volcanic Pistol carries a Minie or conical ball, in a rifle barrel, and will put it through a three inch plank at a distance of ninety rods. The receiving tube will hold ten ball cartridges, which may be deposited in two seconds of time. The pistol may be discharged thirty times in fifty seconds! It is so contrived that it is not liable to accidental discharge. There is no priming — no caps — and therefore no danger to the eyes from any ignition near the breech. Neither is there any recoil, so as to jar the arm or disturb a sure aim. The construction is so simple as to not get out of order even from long use. The powder and ball are enclosed in the same metallic cover so that a person could swim a river with one of these pistols in his belt, without the slightest degree, injuring the powder. In short, the weapon is in all respects one of the most perfect things in the shooting line that we ever took into our hands.

We understand that there is some prospect of the formation of a company here for the manufacture of the arms. We hope, much, that the prospect will be carried out, inasmuch as such a manufactory would add to the general prosperity of the city. The patent gives the exclusive right not merely to manufacture pistols, but rifles and shot guns on the same principle as the pistol. There is already a factory in Norwich whose orders far exceed the capacity of the establishment to supply. The proposition is now to start an establishment that will be sufficiently extensive to turn into market these weapons by the quantity and at less rates than are now necessarily required. Mr. J. Post of this city is the agent for the Patentees, for the sale of the rights of the inventors. — New Haven Palladium."[22]

The hyperbole of this article, common at that time regarding the performance of firearms, is rather apparent, but these exaggerations, which we've already seen in Chapter One with the Smith-Jennings rifle, would soon be repeated with the later Volcanic Repeating Arms Company and the New Haven Arms Company.

For an unknown reason, Wisconsin newspapers seem to have provided the majority of advertisements and announcements promoting the Volcanic. Because the state was then at the edge of the frontier, its citizenry may have been more interested in new firearms. J.W. Storrs, Smith & Wesson's agent in Wisconsin, reported in the *Milwaukee Daily Sentinel*, March, 30, 1855:

"VOLCANIC REPEATING PISTOL — This is a new 'organ of destructiveness,' which rather exceeds in rapidity and efficiency of execution anything we have yet seen. The "ball, powder and priming" are all done up together in a neat looking pill — though like other pills, mighty unpleasant to take — and then dropped, six, ten, or twenty at a time, according to the size of the pistol or rifle, into a tube under and parallel to the barrel, whence, by a single forward motion of the lever, they are transferred, one by one, into the chamber, and discharged at the rate of thirty a minute! The weapon must be seen to be appreciated. Mr. J.W. Storrs, the agent of the patentees (Smith & Wesson) is now in our city, exhibiting this volcanic repeater. It is well worth the attention of sportsmen and others."

Henry J. Nazro began advertising Volcanic arms in May 1855. His advertisement of May 26, 1855, in the *Milwaukee Daily Sentinel* stated:

> ***Volcanic Repeating Pistol*** *— THE SUBSCRIBERS HAVE JUST Received a small lot of the justly celebrated Volcanic Repeating Pistols. Thirty cartridges can be loaded and discharged in 50 seconds. The ammunition is Water-Proof, superior to anything yet out. — H.J. NAZRO & CO.*

The *Milwaukee Daily Free Democrat* and *The Fond du Lac Union* carried this advertisement on May 26 and June 7, 1855:

> ***REVOLVERS*** *Colt's celebrated Revolving Pistol, also Volcanic Repeating Pistols for sale at Nazaro & Co.*[23]

These articles appeared in July 1855, well before the formation of the Volcanic Repeating Arms Company of Norwich, Connecticut. Thus, Smith & Wesson's repeating pistol came to be known as the "Volcanic."

A glowing account of the performance of the "Volcanic Repeating Pistol" was published in the *Scientific American* on November 29, 1854.[24] The rather remarkable but unbelievable testimony printed in this article will be shown in a later on page 122 of Chapter 9.

Perhaps a more fanciful suggestion for the origin of "Volcanic" was suggested by Marcello Terenzi, a noted author and Italian student of antique arms, who suggested that the "Volcanic" was named in honor of Italy's famed Mt. Vesuvius.[25]

GENERAL FEATURES OF THE PISTOLS

All of the Smith & Wesson magazine pistols had an iron frame and used a toggle-link lever action. The barrels were part-octagon, part-round, and although some publications have said they were blued, they were really browned.[26] The frames were made of a rather low-grade, porous iron that did not retain the blued finish well.[27] The frames found today usually have a gray appearance, but a few do retain bluing on the sideplates. The levers and hammers were case-hardened.

The large-frame pistols had a flared butt, while the small-frame pistols had a rounded, bag-shaped grip. The large-frame models were .41 caliber with a capacity of ten rounds, while the small-frame models held six rounds and were .31 caliber. The rear sight was notched and secured to the frame by a screw. There is slight variation in the shape of the front sight, but generally it is a flattened cone. The pistol frames were produced from cast forgings with the other parts being largely interchangeable. However, a certain amount of hand fitting was required

Gold medal presented in 1854 by the Maryland Institute for the Promotion of the Mechanical Arts to Smith & Wesson for their Patent Repeating pistol. Roy Jinks collection

A large-frame Smith & Wesson pistol, serial number "1", with sideplate and grip removed and action open. Note the serial number stamped on the front toggle-link. Alden S. Bradstock, III collection

A large-frame pistol with spring retracted and magazine block open.

and occasionally file marks, especially on the frames, are found. Forging defects on the frames are rather common.

The operation of the pistol was simple: To load the firearm, the knob of the magazine follower was pushed toward the muzzle, compressing the magazine spring until it entered the slot on the underside of the magazine collar. The latter was then rotated counterclockwise exposing the front end of the magazine tube. Cartridges were then inserted into the magazine after which clockwise rotation of the collar permitted the spring to push the magazine follower against the cartridges in the magazine tube, placing a cartridge into the carrier. Moving the finger lever forward elevated the carrier and aligned the cartridge with the chamber and cocked the action. Pulling the lever to the rear moved the bolt forward and pushed the round into the chamber and lowered the carrier. After the pistol had been fired, forward motion of the lever raised the carrier, which ejected the disc base of the cartridge as it moved into position and again cocked the action. Rearward movement of the lever permitted a new cartridge to be chambered, completing the cycle.

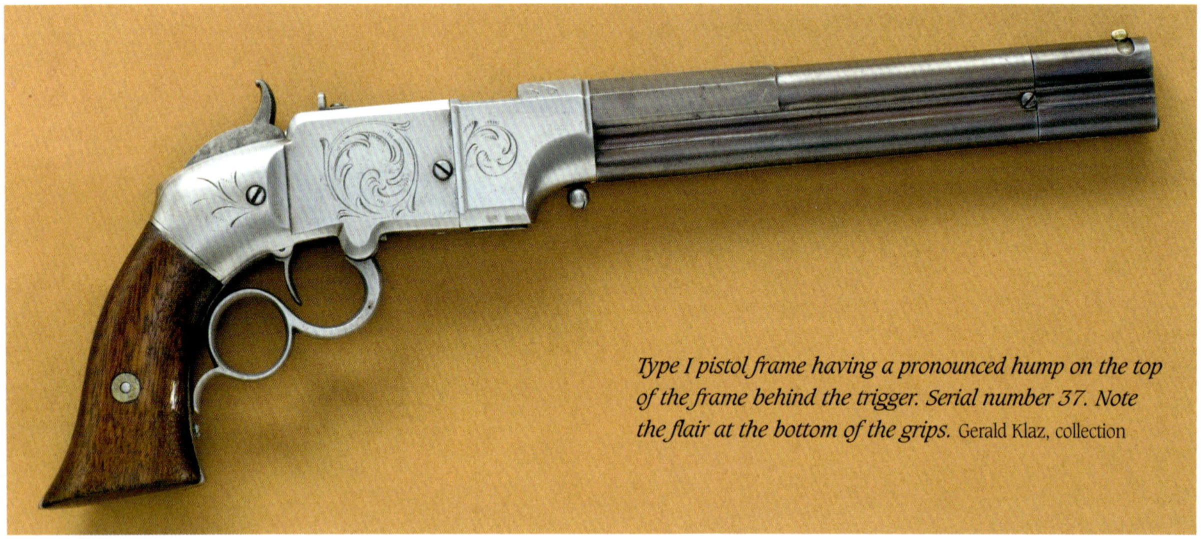

Type I pistol frame having a pronounced hump on the top of the frame behind the trigger. Serial number 37. Note the flair at the bottom of the grips. Gerald Klaz, collection

Problems with the guns were common, and many were returned to the factory for repair. One continuing problem was the frequent misfires. This required the removal of the unfired cartridge from the chamber before a new round could be loaded. Undersized cartridges were a problem; they were forced too far into the chamber and could be neither discharged nor easily extracted.

R. Bruce McDowell nicely summed up the problems with the Smith & Wesson Volcanic ammunition and extraction of the spent rounds in his monograph *Development of the Henry Cartridge* as such:

"The problems with Volcanic cartridges were many. Smith & Wesson had changed their design on the bolt face from the Flobert concept to simply a hook-shaped firing-pin, which was mounted at the center. With the new self-primed rocket ball, the firing-pin had two functions: first, it was a firing-pin which was to strike the primer from a blow from the spring loaded hammer. The idea was good, but variations in cartridge diameters presented still further problems. If the cartridge was too small, it would be forced too far into the barrel chamber and the firing-pin could not come in contact with the priming. If the cartridge was too large, it could not be seated within the chamber and premature discharge often resulted. Further, oversize cartridges would often tear apart upon discharge, with the sides of the powder chamber remaining in the barrel."

From Jinks, *History of Smith & Wesson.*

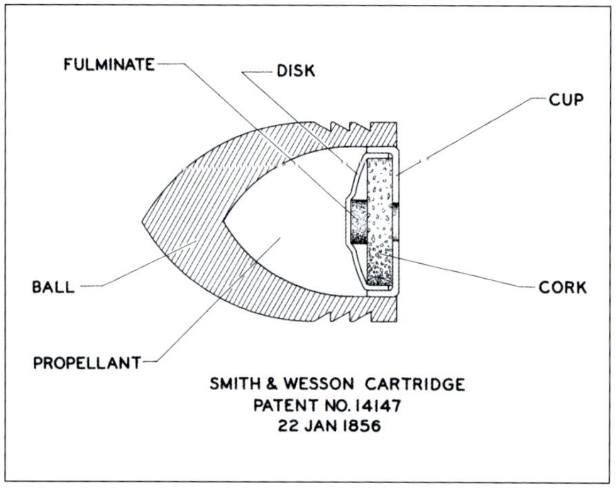

The second function of the hook-shaped firing pin was that of extracting the priming assembly after discharge. The hook was supposed to grab the cork wad firmly enough to extract the entire primer assembly as the bolt assembly was retracted for reloading. This idea worked some of the time, but often it did not, leaving the cork wad or, in the case of a misfire, a portion of the projectile in the chamber.

There were no serious issues with the tubular-magazine repeaters themselves; each was a tribute to mechanical technology. The only problem, and it was an extremely serious problem, was that of the ammunition.[28]

Issues related to the ammunition included lack of striking power, gas leakage at the breech, excessive fouling of the action and the bore because of remnants of the previous round as well as premature detonation of oversized cartridges. What was clearly needed was a better cartridge. Smith & Wesson carried out extensive experimentation, and on May 10, 1853, they filed a patent application for a metallic cartridge, but it was never produced, and the decision was made to use a modified version of Hunt's Rocket Ball cartridge that had a primer in its base. The design of this cartridge, or loaded ball, was submitted in a patent application in 1855, but it was not granted until January 26, 1856.

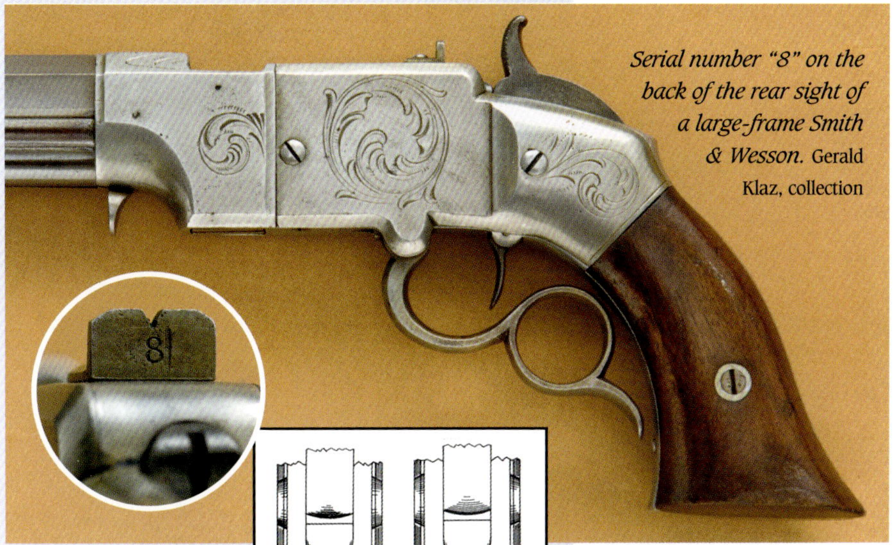

Serial number "8" on the back of the rear sight of a large-frame Smith & Wesson. Gerald Klaz, collection

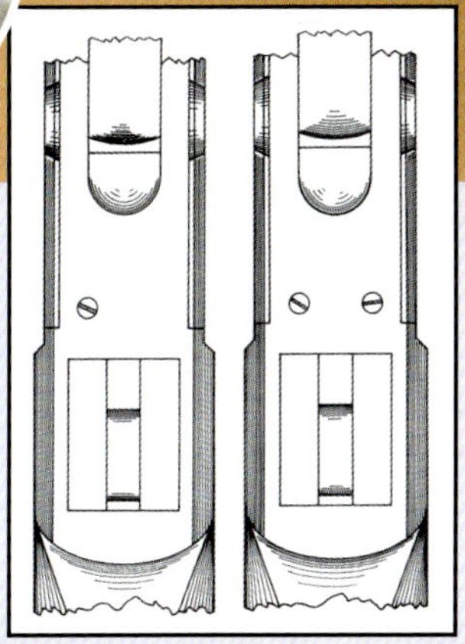

Right: *The early Type I carrier and single-lever spring screw (left) and the two-screw variation (right) found on later Type II and Type III pistols.* From McDowell, *Evolution of the Winchester*

Above: *Type II pistol frame having a pronounced hump on the top of the frame behind the trigger. Serial number 8. Note the flair at the bottom of the grips.* Gerald Klaz, collection

The Type III frame without the hump as shown on the late-production Smith & Wesson pistol-carbine. Gerald Klaz collection

THE LARGE-FRAME SMITH & WESSON REPEATING PISTOL

It is generally thought that the first type of pistol to be produced by Smith & Wesson was a limited production of the large-frame .41 caliber pistol that had been made by Daniel Wesson prior to their partnership. More than likely, this is the reason that these early pistols lack the patent date as part of the barrel markings. The new Smith & Wesson firm was short-lived, and during its brief period of existence, it is believed that fewer than 500 large-frame pistols in .41 caliber were produced.[29]

Type I pistols usually had an 8-inch barrel and, rarely, some were manufactured with a 6-inch barrel. The Type I frame had a pronounced hump on the top of the frame behind the trigger. This more solidly positions the frame in the hand while working the lever, which is seemingly impossible to do with one hand. A similar, but more pronounced configuration is found on some Robbins & Lawrence pepperbox pistols, circa 1851–1854. Both Smith and Wesson were undoubtedly familiar with this design. Type I guns had only one spring screw on the bottom of the frame behind the elevator attaching an internal spring to guide the carrier. They had an uncrowned, flat muzzle, initially with a flat and later a heavy rounded locking spring, attached at its bottom with a screw.

Type II pistols also had a hump on top of the frame. They typically had 8-inch barrels, but several pistols are known with a 6-inch barrel. The primary difference between Type I and Type II pistols was that the latter had a crowned muzzle with a T-bar magazine locking spring secured internally. The later Type II and the Type III pistols had two screws on the bottom of the frame, the additional one securing a second internal spring that held the lever closed. Most of the total production of the large-frame pistols was of these two types.

Type III pistols had a slightly smaller frame, without the hump, and had the more streamlined appearance of the later large brass frame Volcanic Repeating Arms Company pistols. They came in three barrel lengths: 6 inches, 8 inches and 16 inches and those with a 16-inch barrel were provided with a detachable shoulder stock and therefore were pistol-carbines.[30] They also had a crowned muzzle and a T-bar magazine spring. All Type III pistols have the barrel markings of the small-frame pistols including "PATENT FEB. 14, 1854" indicating all Type III pistols, the last of the large-frame production, were produced during 1854 and 1855.[31] Very few were produced.

Serial Numbers — Factory records with the serial numbers of the Smith & Wesson Volcanic arms no longer exist or are known, but some facts are in evidence. The Type I and Type II large-frame pistols were consecutively numbered separately from the small-frame models. The scarce Type III large-frame pistols, produced in 1854 and 1855, were numbered in their own serial number sequence.

Serial number "2" on the frame and on the grip of a large-frame pistol. Courtesy of James Julia, Inc.

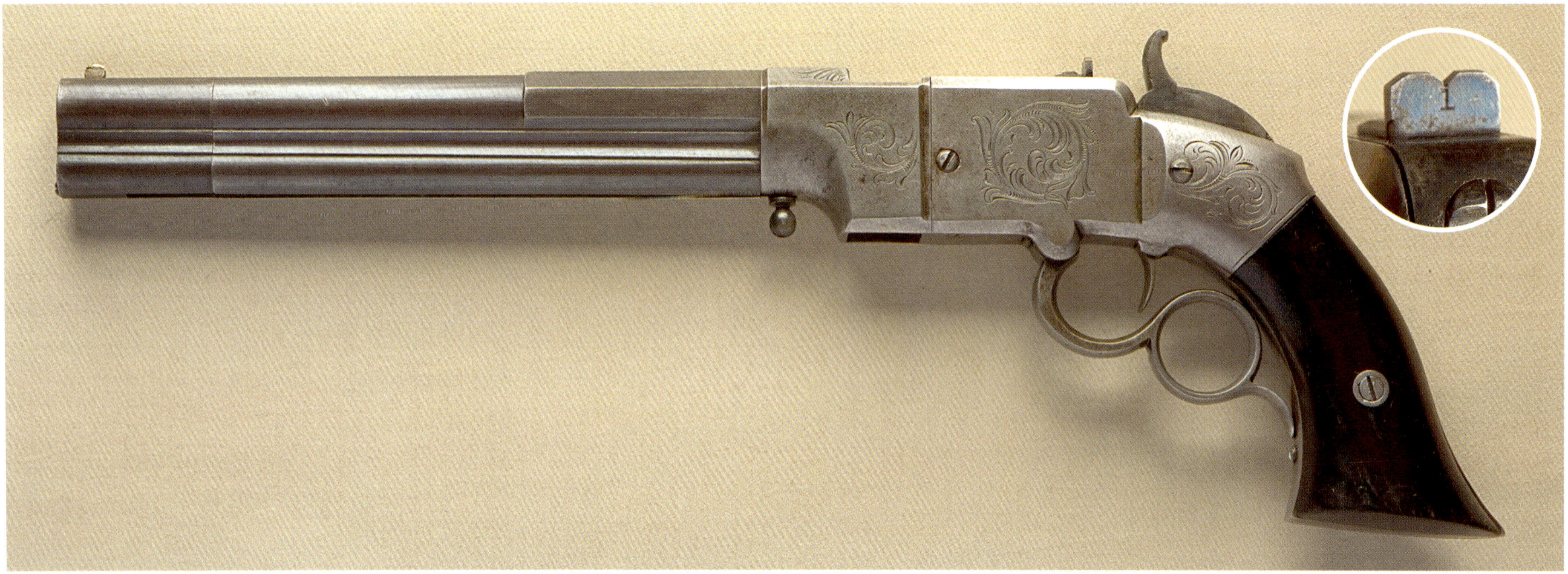

Large-frame Smith & Wesson, serial number "1". The pistol has a lever spur and the early, rounded magazine follower.
Alden S. Bradstock, III collection

Serial numbers are located on the rear surface of the back sight and internally, most often on the left side of the frame. The inside of the grips are often numbered in pencil. There is evidence that more than one serial number sequence was used on the large-frame models. For example, both Type I and Type II pistols are known with serial number "1".[32] The highest large-frame serial number known is 222.

Engraving — Frames, including the top and backstraps, were engraved with a scroll-and-leaf motif. The patterns were of simple design, more attractive than those found on the Smith-Jennings rifles and were meant to increase sales appeal. Coverage was limited in order to quickly cover the surfaces to be engraved and was similar to that found on other firearms of the period. A few pistols were engraved in a more elaborate pattern and at least one had gold inlays. It is unknown whether these were exhibition pieces or given as presentation pistols. As advancements in firearms design occurred, the patterns became more finely executed as seen on some later Henry rifles. Smith & Wesson engraving was simpler on large-frame pistols, with the small-frame guns produced later having somewhat more elaborate engraving. There was some variation in the designs, but generally the same patterns were used. The engravers' names are unknown. Four separate stamps were used in marking the top flats of all Type I and Type II large-frame pistols. The more common early markings had a variable arrangement on the top barrel flat as shown at right. Later, the barrel markings were arranged in an inverted "V". However, the later inverted "V" barrel marking is present on a large-frame pistol with a lever spur, serial number 2. This is likely due to the use of more than one serial number sequence. The markings locations were inconsistent. Some are nearer the breech and others nearer the muzzle. Some lettering was heavily stamped and others quite lightly. Well-worn guns may show barely perceptible stampings and others, none at all. Type I and Type II large-frame pistols were produced before the patent was issued on February 14, 1854, and therefore patent dates were not included in the barrel marking.[33] Type III pistols, the last of the large-frame production, have the barrel markings of the small-frame models, including "PATENT FEB. 14, 1854".

SMITH & WESSON
CAST-STEEL NORWICH, CT. PATENT

SMITH & WESSON
NORWICH, CT.
CAST-STEEL PATENT

SMITH & WESSON
NORWICH, CT. PATENT
CAST STEEL

SMITH & WESSON
CAST-STEEL PATENT
NORWICH, CT.

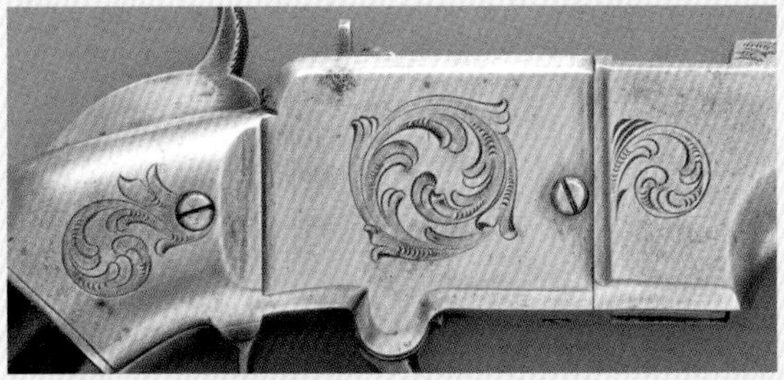

Engraving was simple, and variations were common as shown on these large-frame pistols. Gerald Klaz collection

Below: *Further variations. Most likely these were all engraved by the same individual.* Alden Bradstock, III collection

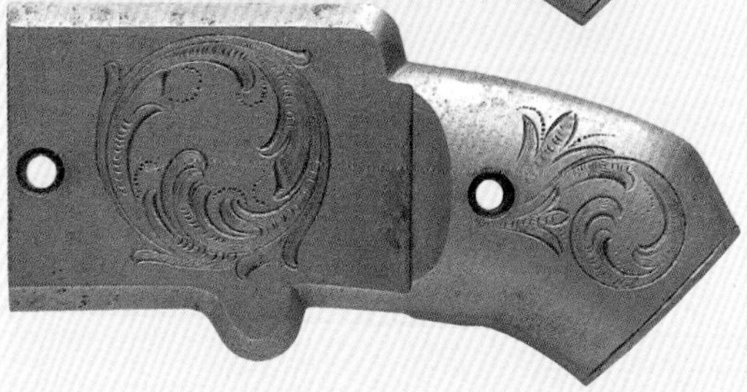

Right: *This large-frame pistol, serial number 47, has more elaborate engraving than found on the standard guns. Note the hunter in the vignette incorporated into the scroll-and-leaf pattern on the right sideplate. A stylized deer is present on the left sideplate. The hammer is engraved, and note "PATENT" stamped on the lever.* Gerald Klaz collection

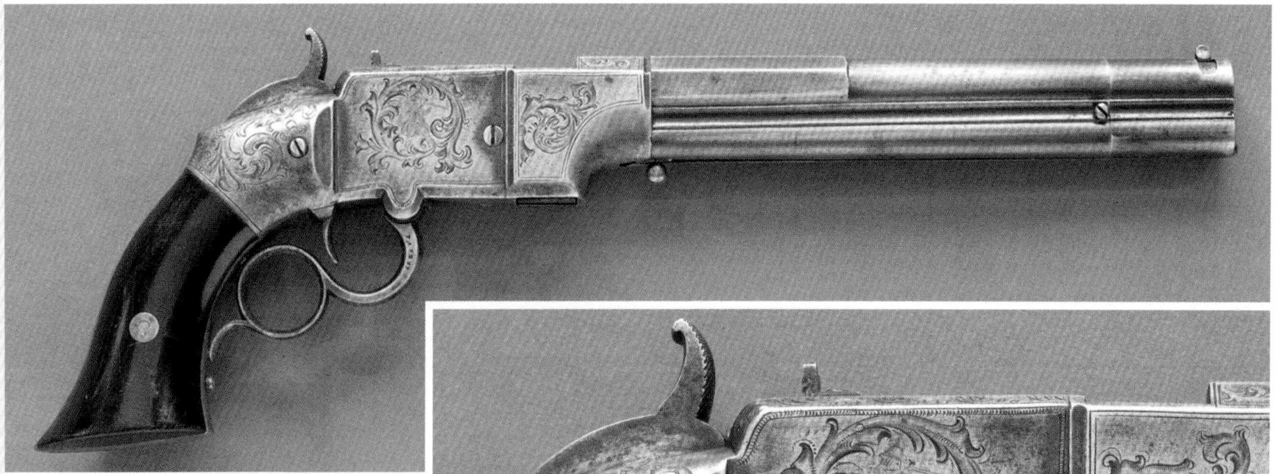

Right: *This large-frame pistol, serial number 81, is fully silver plated as well as being highly engraved and has gold inlays on the barrel and frame along with silver mounts on the rosewood grips. A shield is incorporated into an elaborate scroll pattern on the left sideplate.* Gerald Klaz collection

Below: *Top pistol — The standard length of barrels for the large-frame Type I and Type II pistols was 8 inches. Bottom pistol — Rare Type I Smith & Wesson 6-inch large-frame pistol, serial number 111, formerly in the William Locke collection. Less than ten are known in this barrel length. Note the differences in the magazine block and collar.* Bruce Jordan collection

Type I Smith & Wesson large-frame 6-inch pistol, serial number 27.
Peter DeRose collection

Large-frame pistol, serial number 2, with inverted "V" barrel marking.
Courtesy of James Julia, Inc.

Examples of barrel stamp placement. On the bottom is the most common arrangement. On the top is the later variation referred to as the inverted "V" marking. From McDowell, *Evolution of the Winchester*

Grips — The large-frame pistols had a flared butt and were more hand filling than the bag-shaped grips of the small-frame pistols. They were also made of walnut or rosewood and some had fancy escutcheons surrounding the grip screw. Grips usually had an inside serial number.

Sights — The rear sight was a notched "L"-shaped metal strip attached by a screw. The front sight was a simple flattened cone.

Lever Spurs — Spurs at the lower end of the lever are often found on Type I and Type II pistols, usually below serial number 250. Sometimes the spurs were removed. Spurs were optional on Type III pistols. No Type III pistols with 6-inch barrels had spurs. "PATENT" is stamped on some levers.

Spring Screws — Type I pistols had only one spring screw on the bottom of the frame, to the rear of the carrier, to attach an internal spring guiding the carrier during operation. These generally have serial numbers below 25.[34] Type II and Type III pistols had two internal springs and two screws to hold the finger lever and carrier block in proper position (see illustration on page 23). Variations occur, and some low serial number pistols are known with two screws.

Cartridge Followers — Early Type I pistols had round cartridge followers in the magazine slot to hold pressure against the cartridges therein. Later Type I, Type II and Type III guns had a flattened, "clam shell"-shaped follower. Both types were made of steel. Replacements are often found since the follower was easily broken.

Muzzle and Loading Block Spring Variations — There are two types of muzzle and spring variations that locked the magazine. Type I muzzles were uncrowned and flush with the magazine. The magazine lock was held closed by a flat spring with its lower end attached to the spring chamber by a screw. Early Type II pistols had a rounded, crowned muzzle that extended beyond the magazine lock. On these pistols, the magazine lock was held closed by a coiled, spring-loaded T-bar. Variations are found, and some Type I pistols are known with crowned muzzles. Many of the large-frame Smith & Wesson pistols and the first type small-frame pocket pistols had short muzzle collars that were locked in place by two mounting screws.

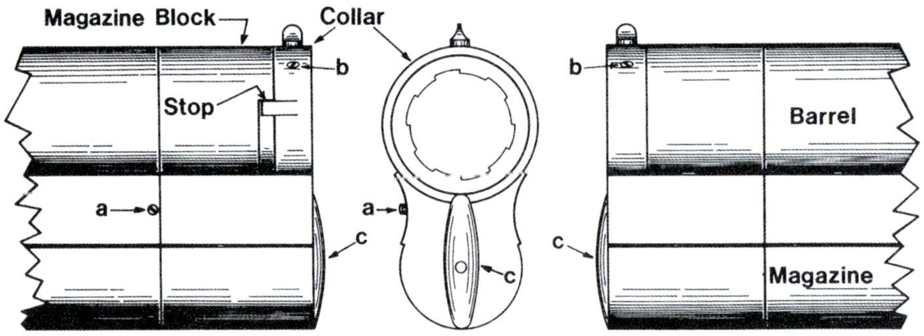

Above: *The early type Smith & Wesson muzzle collar.* From McDowell, *Evolution of the Winchester*

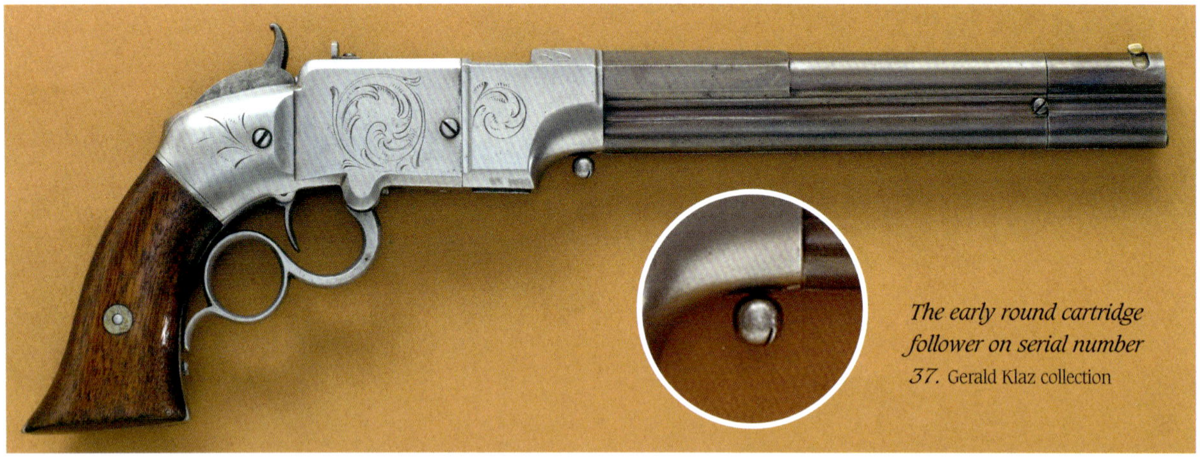

The early round cartridge follower on serial number 37. Gerald Klaz collection

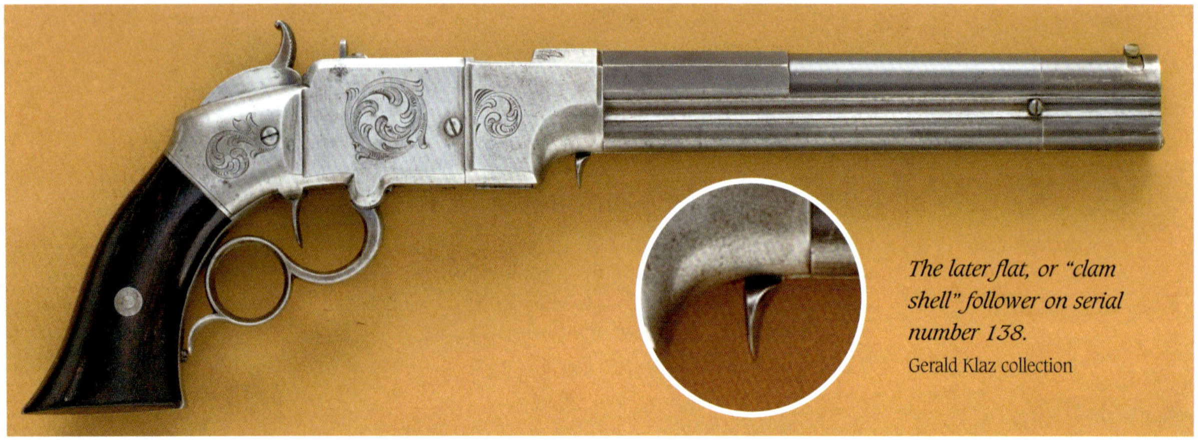

The later flat, or "clam shell" follower on serial number 138. Gerald Klaz collection

Below: Type I muzzle at bottom and Type II muzzle at top. From McDowell's, *Evolution of the Winchester*

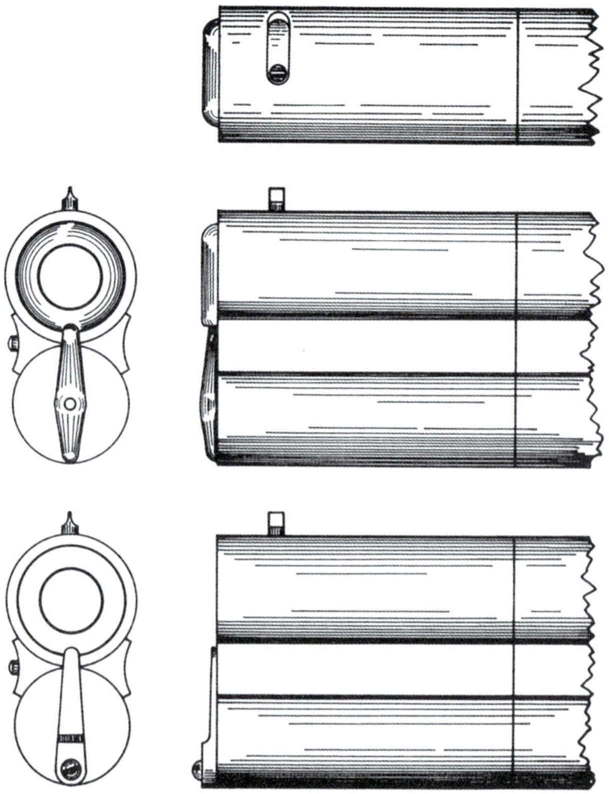

THE SMALL-FRAME SMITH & WESSON REPEATING PISTOL

Below: *Top, an early Type I small-frame Smith & Wesson pistol, serial number 13A, with lever spur. Bottom, later Type II serial number 1, without the spur. This extremely rare pistol is the first in the sequence of guns without a letter suffix. Both with rosewood grips.*
Bruce Jordan collection

Serial number "B 53" stamped inside the frame and grip.

R. Bruce McDowell states, "It is obvious that the small-frame Smith & Wesson Volcanic pistol was designed by Horace Smith. The rounded butt, bag-type grips and small cast iron frames are much like those used on the Allen & Thurber pepperbox pistols of the 1840s and the Percival & Smith magazine pistols manufactured by Horace Smith in 1850. Daniel Wesson's only exposure to the pepperbox design was the .30 caliber, four-shot Leonard pistol manufactured by Robbins & Lawrence."[35]

A few early small-frame Smith & Wesson pistols have been found with the Lewis Jennings patent date, "DEC. 25, 1849", stamped on the barrels. These were made before the Smith & Wesson patent of February 14, 1854, probably as prototypes.[36] It is thought that the small pistols, with their underpowered .31 caliber cartridge, were recognized as less than satisfactory and that larger guns with more suitable ammunition were initially manufactured instead.

The most obvious difference between the small-frame Smith & Wesson magazine pistols and their larger cousins was their size. Their toggle-link action, with minor variations, was the same as found on the large-frame models. They also had an iron frame with a part-octagon, part-round barrel nominally four inches in length (actually 4⅛ inches) with the same blued finish on the frame. The barrels and magazines, again, were browned. Type I small-frame pistols also had a flat muzzle and flat spring, while the Type II pistols had a crowned barrel and was held closed by a more rounded spring-loaded T-bar.

The finger levers and hammers were case-hardened. The grips were rounded and bag-shaped as previously noted, reflecting their Allen & Thurber heritage. They were .31 caliber, and their magazine held six Smith & Wesson No. 1 cartridges. Production of these pistols is estimated to have been 700 by Smith & Wesson, and later, approximately 500 pistols were assembled by the Volcanic Repeating Arms Co. from 1855 to 1857 from parts acquired from Smith & Wesson when that company ceased to exist. In all, about 1,200 pistols were produced.

Small-Frame Serial Numbers — The system for serializing the small-frame pistols is outlined in McDowell's book *Evolution of the Winchester*. The first 100 pistols produced by the firm were numbered consecutively from 1 through 100. After that, Smith & Wesson had an unusual, but logical system of numbering their pistols. This system has allowed a rather precise determination of the numbers produced by them, and later, by the Volcanic Repeating Arms Company. The serial numbers of the small-frame pistols were enumerated as shown in the table on page 32.

The G, H and I series, and higher numbers, were assembled by the new Volcanic Repeating Arms Company from parts acquired from Smith & Wesson after the demise of that company.[37] There was no standard placement of the letters; on some guns, the letter preceded the number, while on others it followed the number or was above or below the number.[38] The serial number is found on the bottom of the carrier lever and internally on the left side of the frame and grips, while the letter, with the number, is found on the inside of the grips.

McDowell tells us that this numbering system was verified by Oliver Winchester himself in a court action brought by the Morse Manufacturing Company:

New York, June 1 1871
Question 9 (by counsel): *"Look at Defendant's Exhibit No. 4 and state whether it has ever been in your possession and from whom you received it, and when?"*
Answer (by Oliver Winchester): *"I recognize this pistol as one I received by express, direct from Mr. John W. Boynton, of Hartford, accompanied by a letter: the date I cannot fix, because I have not that letter with me, but it was since the commencement of this suit. On this receipt, I immediately sought for the manufacturer's registered mark or number, so as to be able to identify it, and approximately to determine the date of manufacture; the mark is C which, by the Smith & Wesson system, would determine it to be the 311th pistol of that construction made by them."*

The serial number of the pistol being examined in court was 11C, and serial number 11C would indeed be the 311th pistol produced.[39] Further, in this same action, Winchester said:

"Among these assets purchased were about 300 pistols of both sizes, finished and unfinished, and parts of some 280 to 300 more pistols, not assembled. The company was occupied in part, nearly or quite a year in overhauling, and repairing and assembling these 600 pistols, more or less and manufacturing a new model and the tools and fixtures for that model."[40]

One might wonder why such a complex system of numbering the small-frame pistols was used when the earlier large-frame guns had a simple sequential numbering system. The reason is unknown, but it has been suggested that the pistols were manufactured by lot with hand fitting by the workers assembling them.[41] This is one of those mysteries that will probably never be solved. Collectors always want answers, but they often forget that things were never done with them in mind.

Barrel Length and Markings — The barrel length of the small-frame model is a nominal 4 inches; the actual measurement is usually ⅛ of an inch longer. The location of the barrel markings on the small-frame pistols is indicated as shown in the table at right. Very few pistols, probably prototypes, carry a "DEC. 25, 1849" patent date, indicating production

	Serial Nos. 1–100B	Serial Nos. 1C and higher
SMITH & WESSON NORWICH, CT.	on the top barrel flat	on the top barrel flat
CAST-STEEL	on the right angle flat	on the right side flat
PATENT FEB 14, 1854	on the left angle flat	on the left side flat

From McDowell, *Evolution of the Winchester*

Above: *The barrel markings on serial number 152 read: "SMITH & WESSON/NORWICH.CT"; Left angle, "PATENT/DEC.25, 1849/FEB. 14, 1854"; Right angle, "CAST-STEEL" (upside down).* Peter DeRose collection

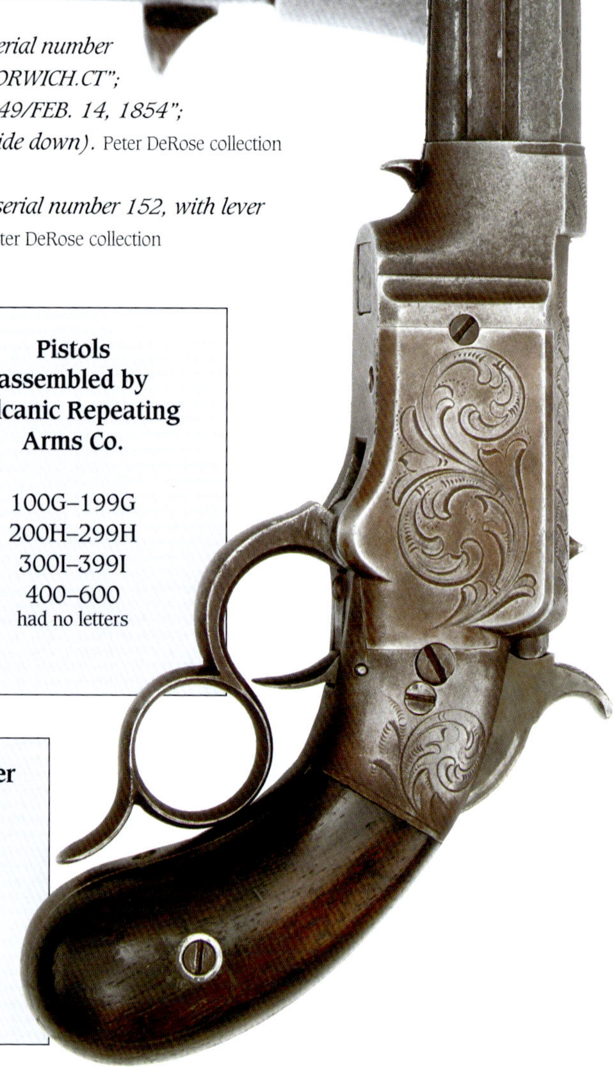

Right: *Type I small-frame pistol, serial number 152, with lever spur and the early flat muzzle.* Peter DeRose collection

Pistols produced by Smith & Wesson	Pistols assembled by Volcanic Repeating Arms Co.
1–100	
1A–100A	100G–199G
1B–100B	200H–299H
1C–100C	300I–399I
1D–100D	400–600 had no letters
1E–100E	
1F–100F	

before the Smith & Wesson patent of 1854. One such pistol in the Nunnemacher Collection at the Milwaukee Public Museum, serial number 8, has a lever spur and a flat muzzle. The illustrated small-frame Smith & Wesson pistol, serial number 152, likely was a transitional pistol since it has both the 1849 and 1854 patent dates stamped on the barrel. Pistols with both a lever spur and an 1849 patent date are quite rare.

Muzzle Collars and Magazine Locks — The first 200 small-frame pistols, serial numbers 1 through 100A, have flat muzzles with the magazine lock secured by either a flat spring or a T-bar. Most pistols in this serial range had a wide muzzle collar. Serial numbers 1B and later have a rounded muzzle crown, a round T-bar magazine retaining spring and a narrower muzzle collar.

Magazine Followers — Magazine followers were made of steel and are of the flat, "clamshell" type rather than the rounded knob found on the earliest Type I large-frame pistols.

Ejection Port Variations — The ejection ports of the small-frame Smith & Wesson pistols are found in two forms. Some are square while others have a half-circle cut-out at the rear. The reason for this variation is uncertain, but perhaps the half-circle cut-out facilitated the ejection of the cork base of the cartridge. This variation is not found on the large-frame pistols.

Above: *Barrel markings on the right, top and left angled flats.* Gerald Klaz collection

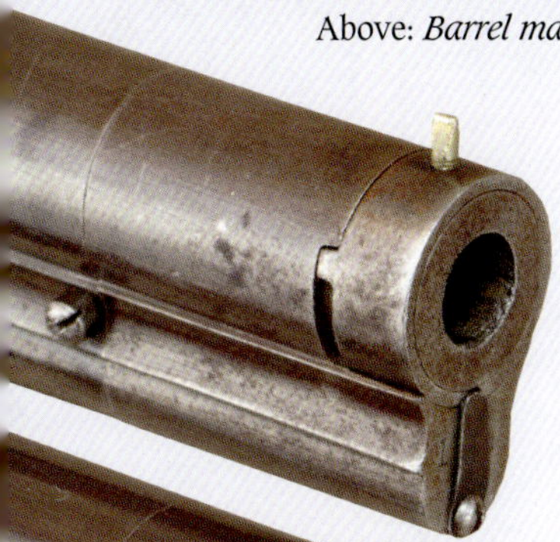

Type I pistol with a flat muzzle, flat spring and wide muzzle collar.

Type II pistol with a heavy rounded T-bar spring and narrow muzzle collar. Bruce Jordan collection

Ejection port variations are found on the small-frame pistols. Left, serial number 19B shows a half-circle cut-out at the rear of the port. Right, serial number 52F has a square cut-out. Donovan Williamson collection

Type II small-frame pistol, serial number E61, with the flat magazine follower. Stephen Rutter collection

Firing Pin and Extractor — The Smith & Wesson pistols, both large and small frame, had a replaceable nipple or firing pin in the center of the bolt face for the self-primed Rocket Ball ammunition. All pistols had a small hook at the front for extraction of the cartridge's priming device after firing. It often failed, leaving the disc in the chamber. Failure of the firing pin extractor was a source of frequent complaint, and the firing pin itself sometimes failed due of erosion from continued exposure to the fulminate. One small-frame pistol, serial number D62, was made with a modified firing pin, which was a V-shaped wedge situated vertically across the bolt face. This type of firing pin was mentioned by Oliver Winchester as having been designed by John Vichers in July of 1856. However, it was not successful and was never placed in production.[42] As discussed later, in Chapter Four, William Hicks continued his attempts to improve the firing pin and extractor at the Volcanic Repeating Arms Company.

Grips — Grips were usually made of rosewood or walnut and were bag-shaped. Serial numbers are found on the inside of the grips, stamped or in pencil.

Levers — Lever spurs were not common. Any small-frame pistol with a lever spur is quite rare. A few pistols from serial number 1 to 100B were produced with spurs, but no pistols with serial numbers higher than 100B are known.[43]

Retaining Springs — Serial numbers 1 through 100B have a single screw on the bottom of the frame holding the carrier lever in position. Serial numbers 1C and higher have an internal spring performing this function with no screw hole on the bottom of the frame. Some higher serial numbers have had the original screw hole filled.[44]

Sights — The sights found on these pistols are quite simple with little variation. The rear sight is a shallow notch at the top of a flattened cylinder inlet into the top of the receiver. The front is a small blade or flattened cone.

Above, left: serial number 95 has one screw; center: serial number B94 has the screw hole plugged; right: serial number I316 has no screw hole. Donovan Williamson collection

A SMITH & WESSON PROTOTYPE .22 CALIBER PISTOL

The .22 caliber cartridge used in Flobert's pistolet de salon re-emerged in a Smith & Wesson prototype model for a .22 caliber pistol. The serial number 233H is found on the right side of the frame and under the grips. The G, H and I series of pistols were assembled by the Volcanic Repeating Arms Company from parts acquired through the reorganization. Whether this prototype was developed by Smith & Wesson themselves or by the new company is uncertain. The 5⅛-inch barrel and magazine assembly is unusual with the magazine block located several inches from the muzzle. The pistol is in the white, never having been blued or browned.

This prototype .22 caliber pistol, serial no. 233H, is marked "SMITH & WESSON" on the top flat of the barrel. The left angled flat has "PATENT FEB. 14, 1854" and the right barrel flat is stamped "CAST STEEL". A modern .22 caliber CB cap fits the chamber perfectly. Note the variation in the carrier. Donovan Williamson collection

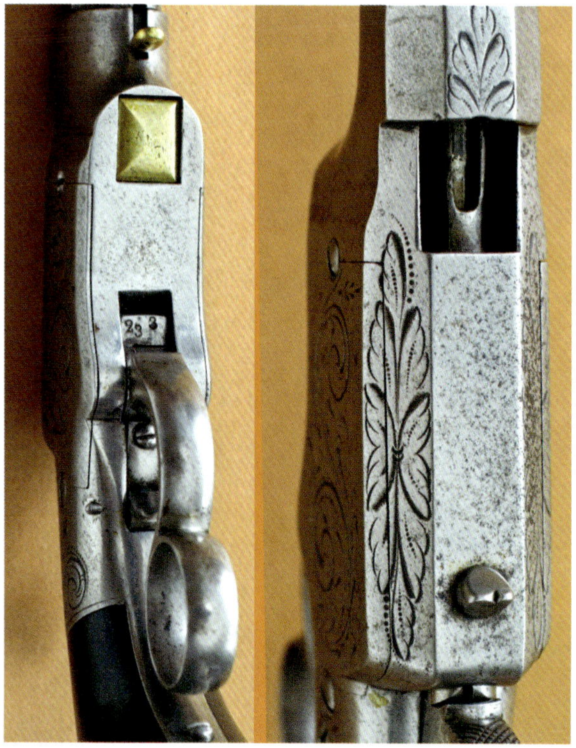

THE SMITH & WESSON PISTOL-CARBINES

It is believed that, for commercial purposes, production at the Smith & Wesson shop in Norwich, Connecticut, was confined to repeating pistols. However, the fact that a few shoulder-type arms were made is evidenced by the pistol-carbines shown here and the rifles to follow.[45]

Primarily noted for the production of the large-frame and small-frame pistols, the company produced a few 16-inch pistol-carbines with detachable stocks, ca.1854–1855. Six were made and two are known to exist.[46] They were

The pistol-carbines were some of the last Type III large-frame firearms produced, circa 1854–1855.
Gerald Klaz collection

made with an iron frame, which lacked the hump at the top of the backstrap and, therefore, had the configuration of the Type III large-frame pistol. This, and the fact that the barrel is marked "SMITH & WESSON/PAT. FEB. 14, 1854/CAST STEEL", indicates that the firearm was of later manufacture and some of the last produced by the company.

This pistol-carbine is a rare jewel that most likely was a Smith & Wesson factory display piece. Its silver-plated frame is profusely engraved in leaves and scrolls. The frame and the top and angled flats forward of the receiver are also engraved. Both the pistol grips and shoulder stock are fashioned of fine rosewood. There is no serial number on the example shown. It is doubtful that it was ever destined for sale.

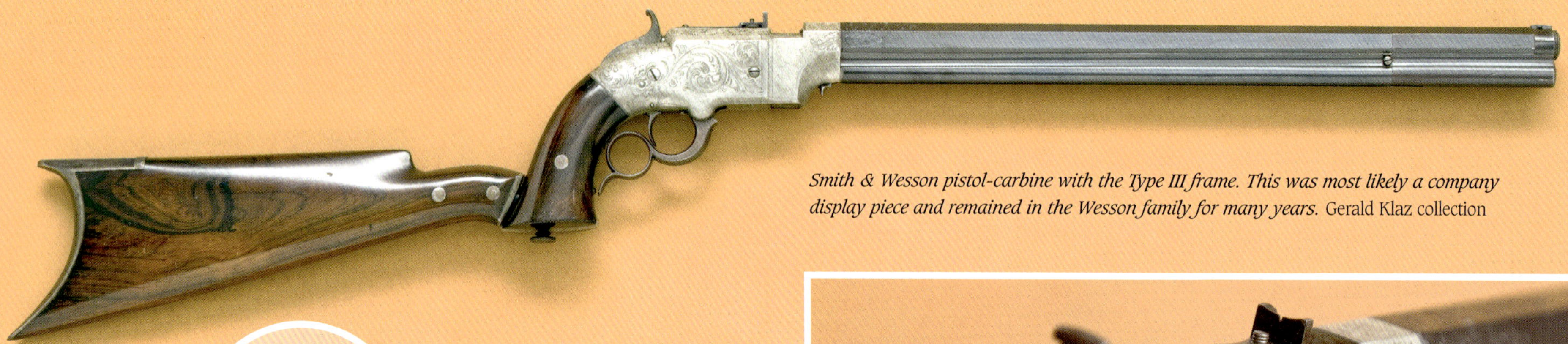

Smith & Wesson pistol-carbine with the Type III frame. This was most likely a company display piece and remained in the Wesson family for many years. Gerald Klaz collection

The pistol-carbine shown here, along with the lever-action rifle pictured at right, were both obtained from the Wesson family by Abercrombie & Fitch, a sporting goods dealer in New York City. They were later acquired by Alan Kelley, a prominent gun collector and dealer from New Haven, Connecticut. Both were then acquired by the present owner.[47]

This Smith & Wesson pistol-carbine is less highly engraved than the other more elaborately engraved presentation or exhibition gun shown. Perhaps this firearm was an early production piece and the company chose to make few others. Any Smith & Wesson pistol-carbine is very rare.

A very rare Smith & Wesson pistol-carbine, serial number 2. Note the quality of the engraving. The barrel address is simply "SMITH & WESSON/PATENT/FEB 14, 1854". The method of attachment is the same as seen on later pistol-carbines made by the Volcanic Repeating Arms Company. Peter DeRose collection

THE SMITH & WESSON LEVER-ACTION REPEATING RIFLES

There are two early advertisements which indicate that Smith & Wesson produced rifles, but it is believed that less than ten were produced.[48] Only a few are known today. These Smith & Wesson repeating rifles were chambered for self-contained metallic-cased cartridges. Since the improved cartridges of U.S. Patent No. 11,496 were never produced for the pistols, it is unlikely that they were produced for the rifles alone. The necessary machinery for manufacturing such cartridges did not exist, and only a few prototype cartridges were made by Smith & Wesson.[49]

Smith & Wesson lever-action rifle with stock inlay and engraved buttplate. Gerald Klaz collection

Facing page: *The integral lever and triggerguard closed and open.* Gerald Klaz collection

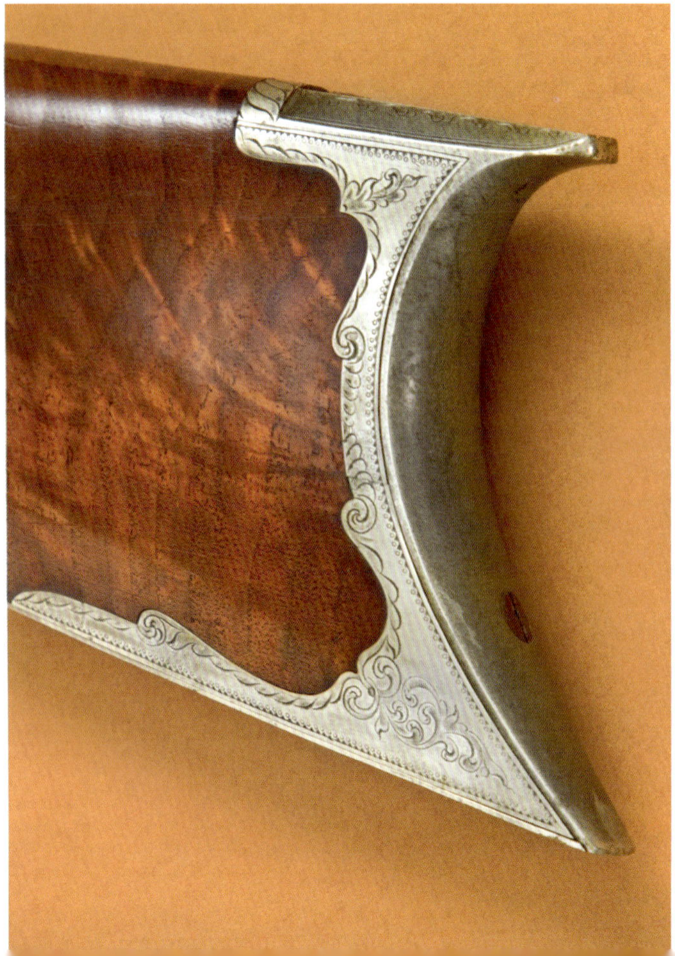

This is one of only two privately owned Smith & Wesson rifles. It has no serial number or barrel marking. It is believed to have been produced at the same time as the pistol-carbine shown earlier and was likely also a display piece. The rifle appears to have never been fired.

The finely engraved iron frame is nickel plated with a triggerguard that is integral with the frame and enclosed by the lever. The part-octagon, part-round barrel is 23 inches in length. Both the barrel and magazine are blued. The stock is of finely figured walnut and has a German silver plaque inlaid on the left side of the butt, engraved with two hands clasped in friendship. The buttplate, also engraved, has German silver inlays as well. More than likely, this was a factory exhibition piece, never destined for sale and never fired. It was manufactured circa 1854–1855. The frame is beautifully engraved as are the top and bottom and the tangs. There is no patent or date marking on the barrel. The front sight is of an unusual, hooded form.

Another Smith & Wesson rifle is in the collection of the Cody Firearms Museum at the Buffalo Bill Historical Center in Cody, Wyoming. It is not as finely finished but is still thought to have been a promotional or exhibition piece. This rifle was probably produced in 1854.[50] The top of the barrel is marked:

"SMITH & WESSON NORWICH, CT. CAST-STEEL PATENT FEBRUARY 14, 1854."

Serial number "8" is marked on the buttstock, buttstock screws, inside of both sideplates and the left side of the frame under the sideplate. The number "1" is found on the left side of the frame and toggle-link and "2" is present on the right side of the frame and toggle-link. These are believed to be assembly numbers. The rifle has a 23-inch barrel, which is octagonal at the breech for six inches. It is chambered for a metallic cartridge of approximately .54 rimfire caliber, probably of the type patented by Smith & Wesson on August 8, 1854. The receiver is bright, and the barrel and magazine assembly are brown. The barrel has the magazine soldered to the underside with a reinforcing rib on both sides. There is a hook-type extractor in the top of the breech bolt and double firing pins, one on each side of the bolt face.

The buttstock is of fancy grained walnut with a German silver inlay on the left side of the stock that is scroll engraved as are the frame, buttplate and lever. The latter has an engraved serpent-head motif. The latch for the finger lever precedes by years the one found on the later model Henry rifle. There is no half-cock notch on the hammer.

These two rifles are early interesting examples of metallic cartridge arms and indicate that even at an early period, Smith & Wesson were aware of the difficulties presented by the loaded ball.

An experimental lever action repeating rifle, with a differently shaped frame, is in the Nunnemacher Collection at the Milwaukee Public Museum.

The last known rifle, in .50 caliber, not illustrated, is in the patent model collection of the Smithsonian Institution.

Ultimately, the demise of the Smith & Wesson Company was due to poor sales and overcapitalization in equipment. It ceased as an entity in July 1855 with the establishment of the Volcanic Repeating Arms Company.

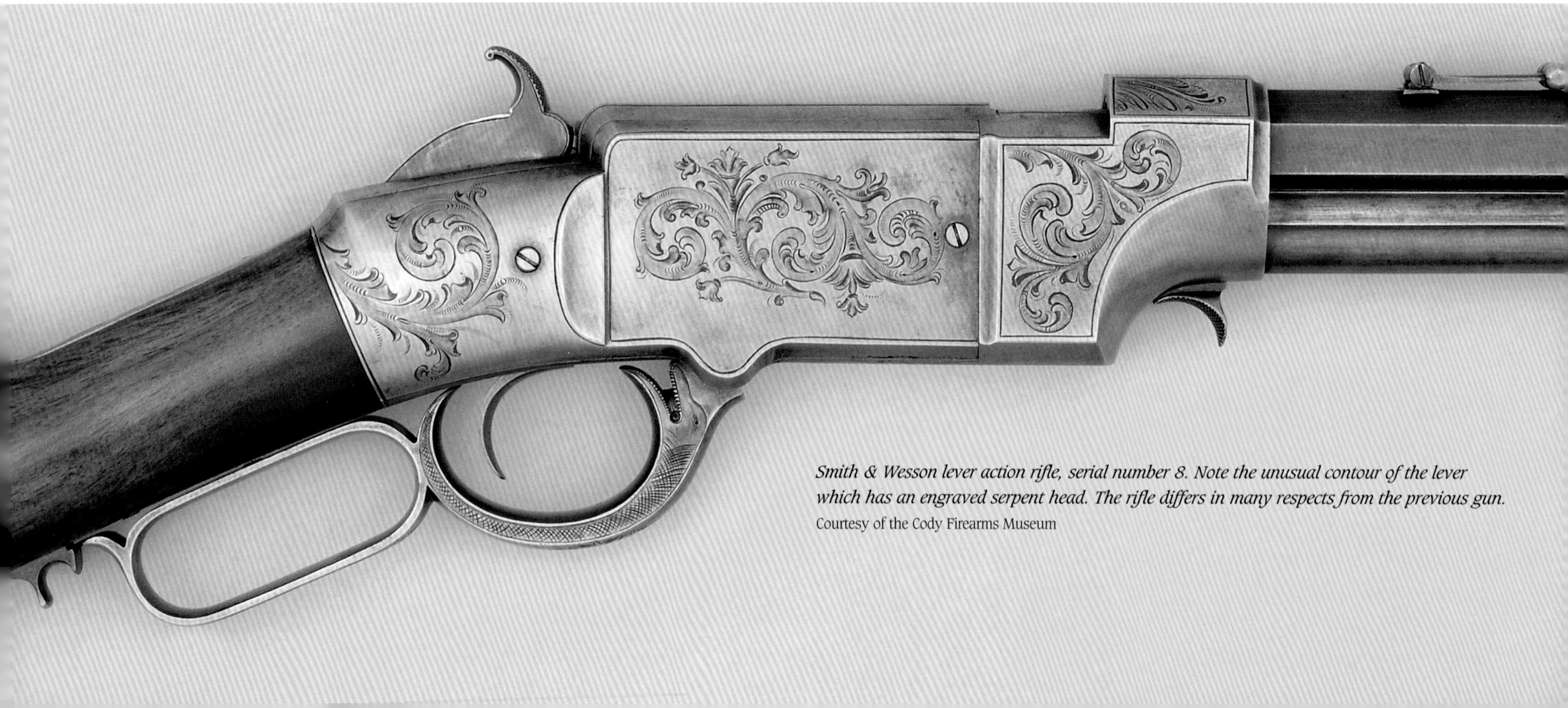

Smith & Wesson lever action rifle, serial number 8. Note the unusual contour of the lever which has an engraved serpent head. The rifle differs in many respects from the previous gun.
Courtesy of the Cody Firearms Museum

Experimental Smith & Wesson rifle, .54 caliber.
The frame has no removable sideplates but is engraved.
Courtesy of the Milwaukee Public Museum,
Nunnemacher Collection

A NEW SMITH & WESSON PARTNERSHIP

After being in existence for only one year, the failure of the Smith & Wesson Company led to a new beginning for Horace Smith and Daniel Wesson. The partners made a decision to sell their assets to a group of New Haven and New York investors. This new company was the Volcanic Repeating Arms Company. Horace Smith made the decision to withdraw immediately from active participation in the new corporation and returned to Springfield, Massachusetts, to work with his brother-in-law. Daniel Wesson was to remain as plant manager for only eight more months, resigning on February 11, 1856.[51]

Daniel B. Wesson continued to work on the metallic cartridge, and in August of 1856, he completed the construction of a wooden model revolver. In November of 1856, Smith and Wesson entered into an agreement with Rollin White giving them an exclusive license to manufacture a pistol having a bored-through cylinder.[52] Thinking this had more promise, they sold their interests and patents in the repeating arms in order to devote their time to the new revolver project.[53] Shortly thereafter, they entered into their second partnership, the Smith & Wesson Revolver Factory of Springfield, Massachusetts, and began preparations for the manufacture of a new revolver. Their First Model seven-shot revolver in .22 short rimfire caliber was introduced in 1857.[54]

The pistols were primarily used for self-protection and were the first firearms produced by a company which was destined to become one of the greatest firearms manufacturers in the United States.

Volcanic Navy, serial number 331. Bruce Jordan collection

chapter three

Volcanic Ammunition

The history of the development of the Volcanic firearms parallels the history of the ammunition that was produced for these guns, a history fraught with unrealized expectations, poor performance and, ultimately, failure. The story begins with the development of Walter Hunt's Rocket Ball, a loaded bullet, U.S. Patent No. 5,701, issued on August 10, 1848. The tale winds its way through the largely unsuccessful early attempts to develop a self-contained metallic cartridge for use in the Smith & Wesson pistols and ends with the final form of the Volcanic cartridge.

Crittenden & Tibbals of South Coventry, Connecticut, was the leading cartridge maker in the United States during the years that the Volition Repeater, and the later Jennings, Smith-Jennings and Smith & Wesson lever-action pistols were being produced.[1] The first sale record found was to Samuel Colt for percussion caps dating February 5, 1849, through June 15, 1849, for 82,000 caps at a cost of $54. No ledger entries or other records survive that mention their association with companies producing the Volcanic firearms.[2] They are, however, thought to have been the first manufacturer of ammunition for the Smith & Wesson Volcanic pistols, and they also later made some of the cartridges for the Volcanic Repeating Arms Company.

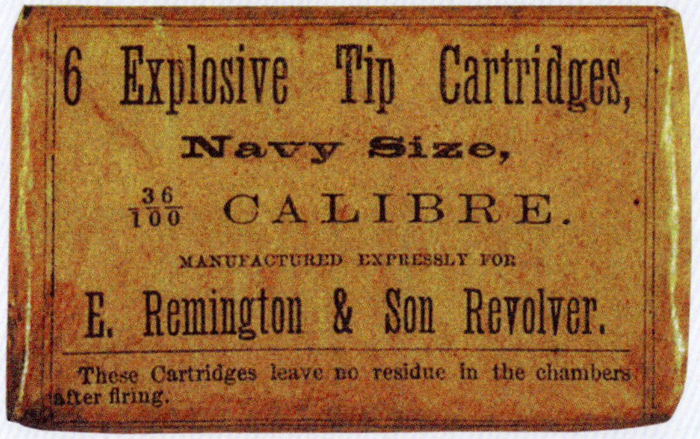

Cartridges manufactured by Wm. Tibbals.
Courtesy of Jack Malloy

RALPH CRITTENDEN was born in Portland, Connecticut, in January of 1820. As a young boy, his family moved to South Coventry, Connecticut. After serving as an apprentice in a machinist's shop for some years, he joined with William Tibbals in the manufacturing of percussion caps. Though he only filed an application for a U.S. Patent, he is said to be the first to put into practical operation a combination of the percussion cap and the rifle ball, a combination which later developed into the more modern cartridges. He later failed to follow through in the process of patenting his invention.[3] He represented South Coventry in the State Legislature of 1856, 1863 and 1865. After selling the firm to Schuyler, Hartley & Graham of New York City, he moved to Hartford where he later died on July 17, 1901.[4]

Hunt Rocket Balls. Note the "U.S. PATENT 1848" stamp on the specimen on the left. Stephen Rutter collection

WILLIAM TIBBALS was born in South Coventry, Connecticut, in December of 1823. He partnered with Ralph Crittenden to form their very successful company. After the Civil War, following the sale of their firm, Tibbals stayed in South Coventry for a few years and later moved to Hartford where he died on September 17, 1890.[5] He remained interested in cartridge development and was granted six U.S. Patents between 1865 and 1869, mainly for variations of cartridge anvils. Among the witnesses for these U.S. Patents was Ralph Crittenden. Tibbals himself also made cartridges, but exactly when is unknown. Some original packets have survived.

THE HUNT ROCKET BALL

It is difficult to understand how the complex Volition Repeater could have been conceived with such little emphasis given to an effective cartridge.[6] Hunt clearly refused to recognize, or was unaware of, previous self-contained cartridge developments. His self-contained Rocket Ball was a highly inefficient cartridge, which was to be used not only in his new gun but the later Lewis Jennings improvement and the Smith-Jennings repeating rifle. Ultimately, the results were to be disastrous. The Hunt and Jennings repeaters were never produced in any numbers. Many Jennings single-shot breechloaders were converted to single-shot muzzleloaders. The Smith-Jennings repeating rifle was a failure because of the lack of fire power due to escaping gases, and the absence of a priming device within the cartridge itself.[7]

Crittenden & Tibbals had expanded their percussion cap production to include the loading of Hunt's cartridge, based upon his U.S. Patent No. 5,701, issued on August 10, 1848, and reissued on February 26, 1850. (See Appendix I)

The Rocket Ball was a conical lead bullet with a cavity at the base filled with gunpowder. With the bullet in a mold, the gunpowder was pressed into the hollow and closed by a cork disc that had an ignition hole in the center. Though Hunt's U.S. Patent called for a metallic disk, only disks made of cork have been noted. It had no primer and required ignition from an external source — a gravity-fed priming pill from the gun's priming magazine located on the top of the receiver. Detonation by the firing pin caused fire to enter the hole at the base of the cartridge which then ignited the black powder. Some of the later Jennings and Smith-Jennings single-shot breechloading rifles were built to use percussion caps in place of the pill primer mechanism, which, when struck by the modified hammer, ignited the powder. The original Rocket Ball was .54 caliber and varied in weight from 330 to 336 grains. The weight of the black powder charge is unknown, but suffice it to say that the cartridge was significantly underpowered, and the usual muzzleloader, with more powder per weight of ball, far better served the rifleman of the day. The Rocket Ball varied from .553 to .560 in caliber. At least six variations have been identified among surviving specimens.[8] On some of the cartridges "U.S. PATENTED 1848" is stamped on the side.

ROCKET BALL TYPES

Type	Length in Inches	Markings
I	1.05	"U.S. PATENTED 1848" on long axis
II	1.05	no markings
III	1.05	Two shallow grooves near base
IV	.89	"U.S. PATENTED 1848" on long axis
V	.89	no markings
VI	.89	"U.S. PATENTED 1848" around bullet

Walter Hunt gave no consideration to grease grooves and had no knowledge of their value. Richard Lawrence, of the Robbins & Lawrence Company in Windsor, Vermont, had developed the use of lubricating grooves to both improve accuracy and to help prevent barrel fouling. Courtlandt Palmer, the financial backer of the Jennings and Smith-Jennings rifles, had directed Lawrence to demonstrate the new Jennings repeater to a prospective foreign buyer. After almost a full day of inaccurate firing, resulting in an excessively fouled barrel, Lawrence came upon the idea of grease grooves, noting that the breechloaded bullet used no greased patch like a muzzleloader bullet did. He cut several grooves in each of the Rocket Balls to be used during the second day and filled the grooves with tallow. The second demonstration of the Jennings repeater resulted in much greater accuracy with a clean barrel.[9]

Although experimental, his innovation proved to be successful, and some Rocket Balls are found with two grooves near the base for a lubricant. In theory, this also permitted a bullet to more easily enter a fouled chamber. The skirt of the cartridge had to be made thinner to allow expansion to seal the bore against frontward gas escape. Examples of this cartridge are scarce, and no boxes of the cartridge are known to have survived. A modification of the Rocket Ball would later be chosen by Smith & Wesson for use in their lever-action repeating pistols.

THE SMITH & WESSON METALLIC CARTRIDGE EXPERIMENTATIONS

Horace Smith and Daniel Wesson were familiar with the Flobert cartridge that was essentially a small, round ball inserted into the end of a percussion cap. It was developed by Louis Nicholas Auguste Flobert and was designed for indoor target use in a "pistolet de salon" — roughly translated as a "Saloon Pistol" — a practice pistol for Paris gallants for the deadly business of defending one's honor, or that of one's lady.[10] In reality, it was little more than a toy. Wesson had purchased a supply of balls and percussion caps from Crittenden & Tibbals for experimentation and later patented a Flobert-style single-shot pistol in 1851 or 1852.[11]

Smith and Wesson realized that the success of their lever-action repeating pistol depended on a more suitable self-contained metallic cartridge. To this end, they applied for a U.S. Patent for such a cartridge on May 10, 1853. The patent covered the feature of a priming compound coating the inside of the base of the metallic case, separated from the powder by a perforated metallic disc. On top of this disc the powder charge was separated by a solid partition from the ball in the open end of the case. The shallow cup formed between the powder partition, and the ball was filled with tallow. After some delay due to possible infringements on Flobert's cartridge patent, U.S. Patent No.11,496 was finally granted to Smith & Wesson on August 8, 1854.

Smith & Wesson clearly had a metallic cartridge in mind for use in their U.S.

Crittenden & Tibbals Mfg. Co., South Coventry, Connecticut. Date unknown. From Krasnickas, *Cartridges of South Coventry*

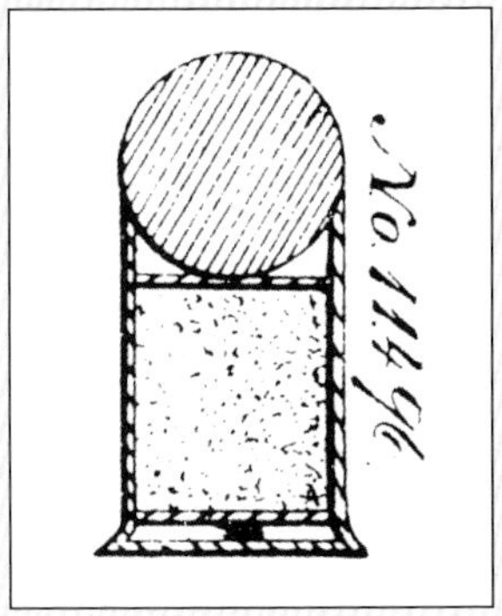

U.S. Patent drawing of the Horace Smith and Daniel B. Wesson improved cartridge, U.S. Patent No. 11,496, issued August 8, 1854. From Jinks, *History of Smith & Wesson*

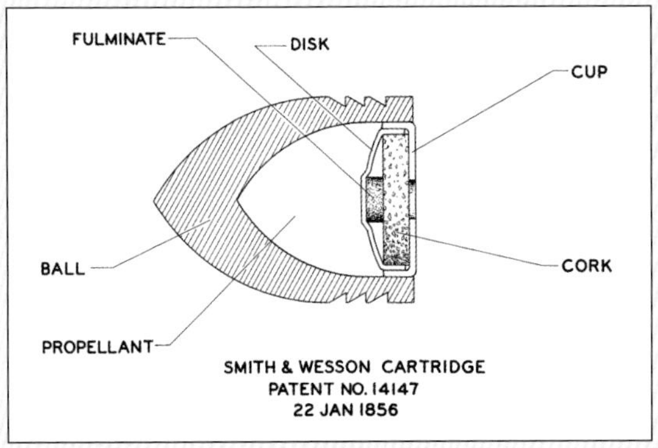

Cross section of the Smith & Wesson Cartridge. From Jinks, *History of Smith & Wesson*

Patent No. 10,535 issued Feb. 14, 1854, for a magazine firearm. The application drawing illustrated the configuration of this type of cartridge, rather than the standard loaded ball cartridge generally associated with this style of firearm, and this claim of novelty may be found:

"...the improvement of making the front end of the piston slide with a dove-tailed recess, or its equivalent, for the purpose of enabling the slide to seize the metal of the cartridge...and so that the refuse metal or cartridge may be withdrawn from the barrel..."

This metallic cartridge was never produced in quantity because, in the early 1850s, little was known about annealing and drawing techniques for copper. Because of the lack of this knowledge, it was virtually impossible to form copper without cracks and tension tears.[12] Some, however, were produced, since several prototype or experimental rifles were chambered for a self-contained, metallic-cased cartridge in .50 caliber. Still faced with the necessity of developing ammunition for their new lever-action pistol, they reverted to a modified form of Hunt's Rocket Ball. The same deficiencies of Hunt's cartridge would continue to plague the modified rounds. They were used in all subsequent Smith & Wesson and Volcanic Repeating Arms Company guns and were known as the Smith & Wesson No.1 or No. 2 Pistol Cartridge. This Smith & Wesson volcanic ammunition is not to be confused with their later No. 1, .22 caliber and No. 2, .32 caliber metallic rimfire cartridges fired in their Model 1 and Model 2 revolvers. The decision to use these underpowered cartridges spelled the demise of not only the first Smith & Wesson Company but also its successor, the Volcanic Repeating Arms Company.

THE SMITH & WESSON "VOLCANIC" CARTRIDGES

The Smith & Wesson No. 1, a .31 caliber cartridge, was made for use in the small-frame pistols; the Smith & Wesson No. 2, a .41 caliber cartridge, was made for use in the large-frame pistols. These cylindro-conoidal-shaped cartridges have also become known as the Volcanic No. 1 Pistol Cartridge and the Volcanic No. 2 Pistol Cartridge. The construction of the two cartridges differed only in caliber, and although some variation is found in the diameter of both the .31 caliber and the .41 caliber cartridges, these caliber designations are generally accepted by collectors.[13] Some later Volcanic cartridges have been found with a .36 caliber bullet diameter. No gun is known, however, which will chamber this cartridge. *The History of Winchester* by Charles T. Haven mentions that, "A .38 caliber Volcanic bullet was recently dissected in the Winchester Laboratories and found to contain a powder charge of 6½ grains plus a small priming pellet of fulminate of mercury and glass in the base of a 100 grain bullet".[14]

As originally made, the bullet was first charged with black powder, then an iron anvil disc was placed, followed by a fulminate priming pill over which a cork base was placed. The bullet was closed at the rear by pressing the lead inward against the cork. Later, a perforated metal cap was sometimes used to cover the cork.[15] After assembly, the ball was grooved and greased, swaged and varnished.

Oliver F. Winchester gives us a much more detailed explanation of the construction of the cartridge in court testimony noted in Appendix II:

"Q4. At the time that the Volcanic Arms Company made the purchase from Smith & Wesson, as mentioned by you, did the company receive any balls from Smith & Wesson; and if yes, what kind or kinds?

A. *They did; of two kinds. I will state that Daniel B. Wesson was one of the directors, appointed at the first meeting for the appointment of directors of the Volcanic Repeating Arms Company. With the purchase from Smith & Wesson, came a small quantity of balls of two kinds; one kind had a primer at the base...formed by a metal disc resting upon the powder, the fulminate pill resting on the centre of that disc, and held in place by a thin circular disc of cork, with the diameter designed to fit tight into the base of the ball; the whole thing being held in place by pressing the lead of the ball snugly around, and slightly over the edge of the cork...The other kind were substantially of the same construction, except that the parts of the primer, the disc, fulminate, and cork, were enclosed in a brass cup having a small circular hole through the closed end, which was pressed down and doubled over on the inner edge, so as to bind the whole together in one piece. Mr. Wesson as a director of the company, called the attention of the board especially to this last named ball* [to] *explain the object of its construction, the difficulties it overcame, and advised immediate arrangements for procuring a supply... Such arrangements were immediately made, in consequence of Mr. Wesson's recommendation, with Crittenden & Tibbals of South Coventry for a supply. It was the cartridge thus adapted by the company, and no other was used by them to my knowledge, during its existence."*

Earlier mention was made of the grooves created near the base of the bullet in some of the Rocket Ball cartridges. These grooves were continued in the construction of the Volcanic cartridges where two styles of grooves are found on the conventional Volcanic cartridges. One style, which is suspected as being the earlier of the two, has five or six fine grooves. The other style, thought to have been used later, has only three or four rather coarse grooves. The reason the finely grooved type is suspected of being the earlier is that almost all finely grooved specimens examined have iron anvils, while the majority of the coarsely grooved Volcanic cartridges examined employ a brass anvil.[16] The Volcanic cartridges were manufactured with the variations of the bases:

	No. 1	No. 2
Before Aug. 1855	cork base	cork base
Aug. to Dec. 1855	copper cap	copper cap
Jan. 1856 to 1860	brass cap	brass cap[17]

Copper caps proved unsatisfactory because of faulty extraction and are very rare. Thereafter, brass was used to facilitate extraction.[18] The brass anvil was a decided improvement over the iron anvil because iron has a habit of reacting unfavorably with fulminate over time, rendering the primer inactive. The manufacturers of the Volcanic cartridge must have discovered this and switched from iron to brass shortly after they instituted the coarse grooving. Just when this happened is a matter of conjecture, but a good guess would be in the late 1850s. The United States Ordnance Department made the same discovery several years after the Civil War when they permanently discontinued the manufacture of all anvil types made of iron.[19]

It has been said that the poor performance of the .31 caliber round led Smith & Wesson to develop a larger .41 caliber cartridge, which was used in the large-frame Smith & Wesson pistols and, with modifications of the priming mechanism, in the large-frame pistols of the Volcanic Repeating Arms Company and the later large-frame and small-frame pistols and carbines of the New Haven Arms Company. The No. 2 Pistol Cartridge had a black powder charge of 6.5 grains in a 100-grain projectile, which generated a

Above: *Left to right — No. 1 Pistol Cartridges: .31 caliber, cork base, six grooves; .31 caliber brass base, five grooves. No. 2 Pistol Cartridges — .41 caliber, brass base, five grooves; .41 caliber, brass base, six grooves; .41 caliber, cork base, six grooves; .41 caliber "dud." Note the impression of the firing pin.* Ray Giles and Donovan Williamson collections

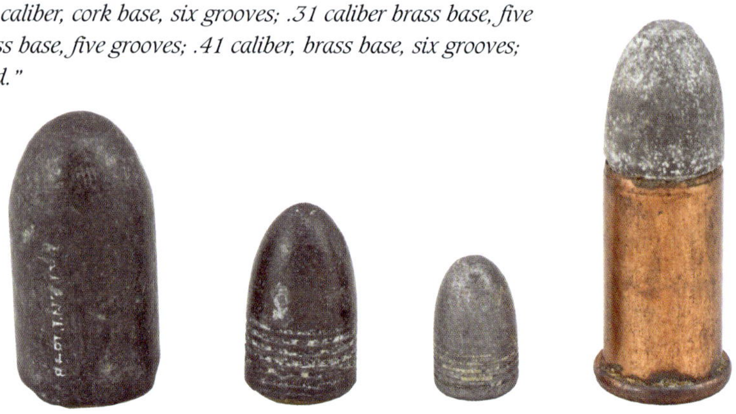

Right: *Left to right — the Hunt rocket ball with "U.S. PATENT" marking, .41 caliber, No. 2 Pistol Cartridge; .31 caliber; No. 1 Pistol Cartridge; an early .44 Henry pointed cartridge with an impressed "H" on the base.* Ray Giles and Donovan Williamson collections

Below: *Possible experimental .54 caliber Volcanic cartridge on left, with a .41 caliber No. 2 Pistol Cartridge and .31 caliber No. 1 Pistol Cartridge.* Ray Giles collection

muzzle velocity of only 260 fps. This was a less-than-ideal performance. Even the short rimfire cartridge of the 1863 Derringer was more powerful. With its 13 grains of powder in a 130-grain projectile, the muzzle velocity of the Derringer was 520 fps.[20]

The performance of the No. 1 Pistol Cartridge was even less satisfactory. One anecdotal story about the performance of the .31 caliber Pocket Pistol was related by Allan Kelley, a well-known gun dealer from New Haven: A despondent gentleman sat on his door stoop one evening after his family had left him alone to run errands. The man decided to depart this world, and proceeded to shoot himself several times in the temple with his pistol, presumably working the lever for each shot. His family returned to find him bloodied but alive with the gun still in his hand. Afterwards, the pistol and remaining ammunition were sold by the family to Kelley.[21]

Crittenden & Tibbals of South Coventry, Connecticut, were the first manufacturers of the .31 caliber and .41 caliber Smith & Wesson Volcanic cartridges. On July 10, 1855, the assets of the partnership of Smith & Wesson became the sole property of the Volcanic Repeating Arms Company. When Oliver F. Winchester bought the assets of the Volcanic Repeating Arms Company Co., he acquired the rights to the Smith & Wesson ammunition patents and all future improvements derived from them.[22] The Ammunition Department of the Volcanic Repeating Arms Company was later established on May 1, 1856, with four female workers, transferred from the Gun Department, working under the direction of M.L. Babcock as foreman. Previously, the ammunition had come from Norwich and South Coventry. By 1856, there were seven women employed in the shop. At the time, Crittenden & Tibbals still provided the primers.[23] On August 26, 1864, Crittenden & Tibbals sold all of their holdings to Jacob R. Schuyler, Marcellus Hartley and Malcolm Graham of New York City, and the firm became known as the Union Metallic Cartridge & Cap Company. A list of the transferred assets included "two sets of machinery for Volcanic Arms."[24]

As long as it remained in stock, the Volcanic ammunition continued to be sold to military goods dealers and gun owners. However, on September 3, 1863, the New Haven Arms Company replied to Tryon & Brothers of Philadelphia: "...we are entirely out of No. 1 Balls and it is uncertain how soon we shall have any more, as the parties who made them have changed their machinery for other purposes. As soon as we can succeed in getting any we will inform you."[25]

PROBLEMS WITH THE CARTRIDGES

With care, comparatively little trouble was experienced with the Volcanic cartridges. There are records of over 100 consecutive shots being fired with a Volcanic carbine before any difficulties were encountered. Such a performance was a wonder. In the day of these arms, when muzzleloaders were standard, B. Tyler Henry, who was probably in charge of production of the Volcanic pistols, testified that he never experienced trouble with misfires in his job of personally test-firing each one made.[26] In a later chapter, we will find that some individuals preferred them to Colt's pistol.

That said, there were some serious drawbacks to the cartridges. While the self-contained ammunition was relatively waterproof and easier to carry than loose powder, balls and caps, these advantages were offset by their many problems. They were underpowered, and the claims of excellent, long-range accuracy were overblown. The small hollowed-out bullet could not hold a sufficient amount of powder to achieve significant velocity. There was always a possibility of the discharge of cartridges in the magazine should undue pressure be exerted on the thinly protected priming charges in the base of each bullet. There was also the usual fault of all breechloading guns before the advent of the metallic cartridge, namely that the bullet had no effective shoulder and was incapable of sealing the breech on detonation, thus causing a leakage of gas.[27] The balls were sometimes oversized and would not enter the chamber but would discharge with pressure from the bolt with unexpected and undesirable consequences. Sometimes an undersized cartridge would be pushed too far into the chamber and would not fire, requiring removal with a rod, as was prominently mentioned on the cartridge box label. Increasing the friction between the cartridge and the walls of the chamber to prevent this resulted in another problem: "thimbling," where the head of the ball would exit the barrel but leave the rear portion of the bullet lodged within the chamber.[28] Another common problem was the incomplete extraction of the priming disc, leaving a portion in the chamber and rendering the arm inoperable since a new round could not be loaded. At least, in comparison with one of Colt's percussion revolvers, if one charge failed, you could go on to the next. Lastly, the priming fulminate often caused corrosion of the firing pin requiring its replacement.

The many problems with the ammunition contributed to poor sales resulting in the bankruptcy of the Volcanic Repeating Arms Company. Later, when the same ammunition was used in the firearms of the New Haven Arms Company, it endangered that company as well. A better cartridge was clearly needed. The ammunition problem would finally be solved when B. Tyler Henry developed his metallic .44 rimfire cartridge, thus assuring the success of the company.

VOLCANIC CARTRIDGE BOXES

The U.S. Patent date of February 26, 1850, is for the reissue of Hunt's U.S. Patent No. 5,701 for "IMPROVEMENT IN METALLIC CARTRIDGES", while August 8, 1854, is the date of Smith and Wesson's "IMPROVEMENT IN CARTRIDGES," U.S. Patent No. 11,496, for the metallic cartridge that was never produced. The gray, non-lacquered tin boxes held 200 Volcanic No. 1 Pistol Cartridges and were probably produced by Crittenden & Tibbals despite the claim of manufacture by Smith & Wesson.

Although the ammunition for these firearms was most likely manufactured by Crittenden & Tibbals, there are no known Volcanic cartridge boxes bearing their name. The cartridges, whether of .31 or .41 caliber, were packed two hundred to the box, and sawdust was sometimes used as filler between the rounds.

The earliest and most commonly found boxes probably produced by the Volcanic Repeating Arms Company were made of black japanned tin, which was then lacquered. The label on this style of box is found in varying shades of off-white or buff and was pasted on the inside of the lid. Later, cardboard boxes were manufactured by the company and are extremely rare, probably due to their fragility and, with the Volcanic firearms becoming obsolete, decreased production. They have a green label pasted on the top of the box. The lettering style and the borders resemble boxes that were later produced for the Henry Rifle.

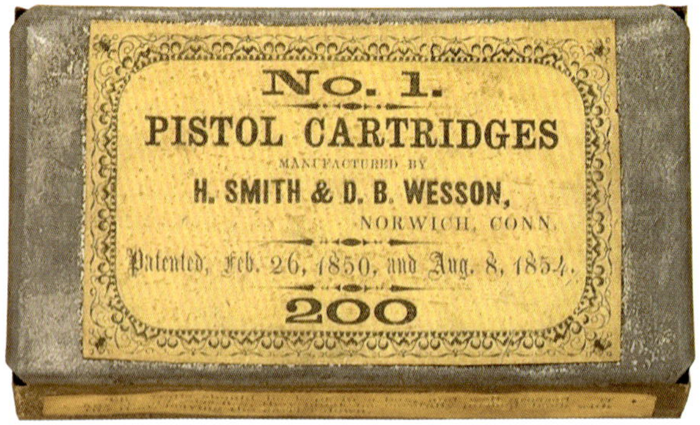

Shown here is a very rare, and unique, box of Smith & Wesson cartridges. Raymond Bentley collection

The label reads:

No. 1.
PISTOL CARTRIDGES
MANUFACTURED BY
H. SMITH & D.B. WESSON,
NORWICH, CONN.
Patented, Feb. 26, 1850, and Aug. 8, 1854.
200

Above: *The earliest box was a lacquered, black japanned tin box with the label inside of the cover.* Alden S. Bradstock, III collection

Below: *Two later cartridge labels with different borders.* Dean Thomas collection

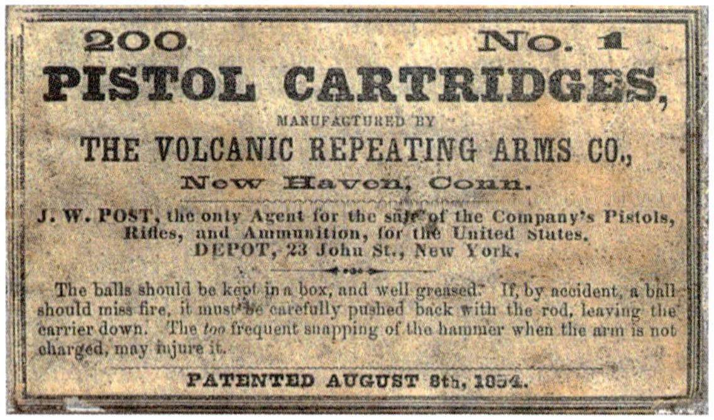

Below: *A slightly later label for a box of No. 2 PISTOL CARTRIDGES produced by the Volcanic Repeating Arms Company. This box includes a more decorative border.*
Dean Thomas collection

Below: *The later, very rare and possibly unique cardboard box of No. 2 PISTOL CARTRIDGES manufactured by the Volcanic Repeating Arms Company with a top label. Note the decorative "Henry-like" border.*
Courtesy Cody Firearms Museum, Buffalo Bill Historical Center

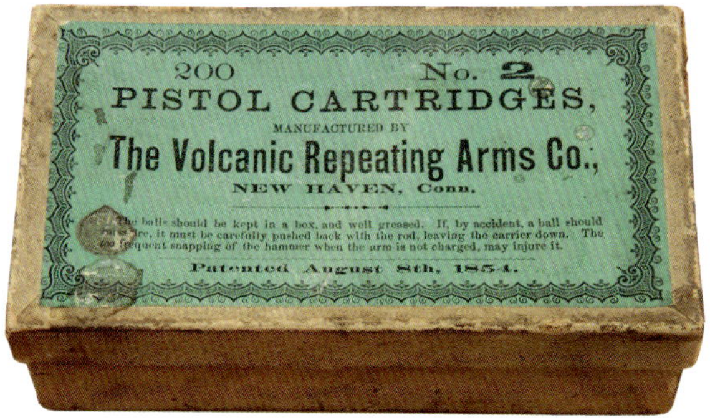

Early boxes of No. 1 and No. 2 PISTOL CARTRIDGES manufactured by the Volcanic Repeating Arms Company list J.W. Post as agent for the company.[29] Later boxes do not include an agent's name. The label included these directions for the use of the cartridges:

"The balls should be kept in a box, and well greased. If, by accident, a ball should misfire it must be carefully pushed back with the rod, leaving the carrier down. The too frequent snapping of the hammer when the arm is not charged may injure it."

The date at the bottom of these boxes, "U.S. PATENTED AUGUST 8, 1854", is not the date of Hunt's Rocket Ball, U.S. Patent of August 10, 1848, or of Smith & Wesson's priming device, U.S. Patent of January 22, 1856, but rather is the date of Smith & Wesson's U.S. Patent for their metallic cartridge. Strangely, that date has nothing to do with Volcanic ammunition. It should read "January 22, 1856", the date of their U.S. Patent No. 14,147 for "IMPROVED PRIMERS FOR CARTRIDGES OF FIREARMS".[30]

Most Volcanic firearms were provided with one or two boxes of ammunition at the time of purchase and more could be procured by their agents. These prices for ammunition were seen on an advertising broadside:

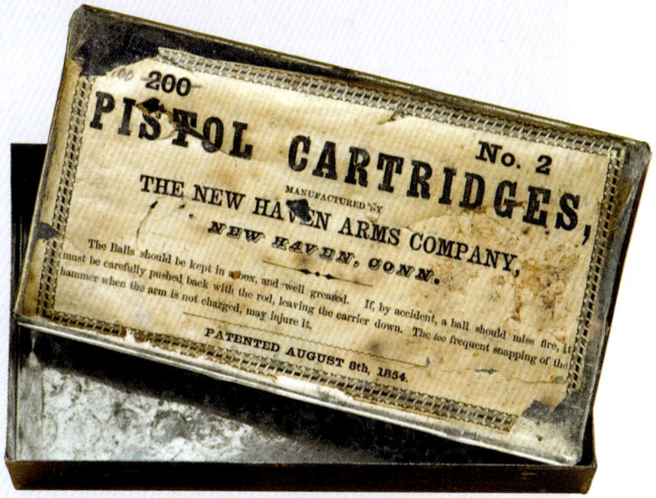

No. 2 PISTOL CARTRIDGES, manufactured by the New Haven Arms Company.
Raymond Bentley collection

No. 1 Balls, 130 to the pound, $10 per thousand
No. 2 Balls, 66 to the pound, $12 per thousand

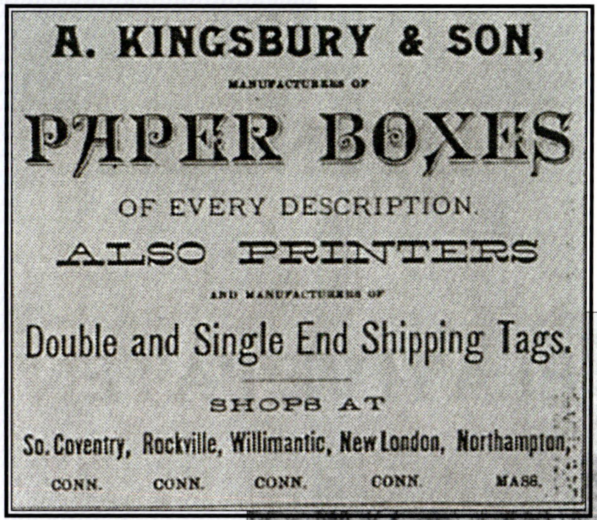

One author has stated that, "The cartridges came in round tins holding 200 each, which sold for $10 and $12 per thousand."[31] Though numerous inquiries have been made, no one seems to have seen a round tin, but a pistol case is known with two round compartments that could have held such tins.[32] The cardboard boxes were most likely produced by the Kingsbury Box Company of South Coventry, which was located very near the Crittenden & Tibbals Company.

Several individuals have related stories of dealers and collectors, who have sold individual cartridges from full boxes and when the box was empty, to the chagrin of collectors, simply discarded the box.[33] This is perhaps one of the major reasons why Smith & Wesson boxes are virtually unknown and Volcanic tin boxes are rare. Boxes of the Hunt Rocket Balls are unknown, but if one is found it would most likely have a Crittenden & Tibbals label.

Right: *The Kingsbury Box Company of South Coventry, Connecticut. Their advertisement is reproduced above.*
From Krasnickas, *Cartridges of South Coventry*

AFTERTHOUGHTS

In the years following production, even though obsolete by any standard of judgment, Volcanic cartridge ammunition was still being sold and used as evidenced by the following letter:

New Haven, Oct. 16/62

Messrs. Merwin & Bray,
New York City.
Gentlemen:
We send you by Express tonight two Henry Rifles & 500 Cartridges;
1 Walsh Revolver & 8 inch Volcanic Pistol repaired and 1000 #2 balls...
New Haven Arms Co.
J.H. Conklin, Secy.[34]

Long after the Volcanic ammunition proved inadequate, the legacy of the Volcanic firearms toggle-link mechanism lived on in the form of the Henry and early Winchester repeating rifles with their improved .44 rimfire metallic cartridge, followed by the further advances in ammunition development with the production of the centerfire cartridge in the 1870s.

chapter four

NEW BEGINNINGS

The Volcanic Repeating Arms Company

Top right corner: *Winchester & Davies New Haven Shirt Manufactory advertising broadside, ca. 1860.*
From Houze, *Winchester Repeating Arms Company, Its History & Development from 1865 to 1981*

Below: *This brick building on Orange Street in New Haven, Connecticut, was the first office of the Volcanic Repeating Arms Company.*

The partnership that Horace Smith and Daniel Wesson formed with Courtlandt Palmer on June 20, 1854, failed in little more than a year due to the lack of the commercial success of their pistols and the poor performance of their ammunition. The Smith & Wesson Company was reorganized as the Volcanic Repeating Arms Company in July of 1855 when it was incorporated under the general laws of the State of Connecticut with the principal office located in New Haven.

The capital stock of $150,000 was divided into 6,000 shares with a par value of $25 per share. The incorporators were Horace Smith and Daniel B. Wesson, both of Norwich, Connecticut, and Courtlandt Palmer of New York City. There were 33 subscribers to the stock, many of who were clock makers.[1] Clock making and the manufacture of carriages were then the principal industries in New Haven. Ultimately, the most important subscriber was Oliver F. Winchester, a shirt manufacturer living at 57 Court Street in New Haven.

OLIVER F. WINCHESTER was born in Boston, Massachusetts, on November 10, 1810, a twin and the youngest of five children. His father died when he was an infant, leaving the family destitute. In his early years, he worked on a farm, later became a carpenter, and after laboring for six years, became a master builder. Leaving Massachusetts for Baltimore, Maryland, he became a successful building contractor but left that field to work in a dry goods store. Soon after, he opened his own store in Baltimore selling men's furnishings. Winchester was a man of considerable intellectual curiosity, and it was apparent that he was destined for success. In 1847, Winchester sold his Baltimore operations and moved to New Haven, Connecticut, to produce and market an improved men's dress shirt for which he had been granted a patent on February 1, 1848. It involved a method for cutting men's shirts to avoid pull on the neckband.

One year later, he partnered with John Davies, a man with great experience in marketing. They expanded production of shirt manufacturing and soon employed over 800 people to cut shirts, with the pieces being sent out to about 5,000 workers to be sewn at home by hand. The firm of Winchester & Davies became very profitable, and Winchester himself became very wealthy. Winchester, soon to be the most famous rifle maker in the world, had bet his shirt on his shirt factory and won. Though having no experience or great knowledge about firearms, Winchester was an astute businessman and thought he saw promise in a new gun company. He invested $2,000 in the stock of the Volcanic Repeating Arms Company. Wisely, however, he remained active in the shirt manufacturing business.

When formed in June of 1855, the Board of Directors of the Volcanic Repeating Arms Company made the following appointments:

President. . . Nelson Gaston of New Haven	Vice President . . . Oliver F. Winchester of New Haven
Secretary. . . Samuel Talcott of New Haven	Plant Manager . . . Daniel B. Wesson of Norwich

Volcanic Repeating Fire-Arms.

Rifles and Pistols carrying from 7 to 30 Balls. Thirty Balls can be loaded and fired in less than one minute.

REPEATING FIRE-ARMS,

AND PATENT LOADED BALL.

Push the spring up in the tube by the knob A, till the top or cap B can be turned to the left, then put the number of cartridges you wish to use in the tube, replace the cap B, when the spring will follow the cartridge down; raise the hammer C, and swing the lever D CLEAR forward, which will elevate the carrier E with a cartridge; pull the lever D CLEAR back, which forces the cartridge into the barrel and braces the breech-pin F, when the arm is in condition for discharge. The balls should never be carried in any position to get the grease wiped off. The arm will never miss fire if these directions are observed.

On July 10, 1855, Courtlandt Palmer, Horace Smith and Daniel B. Wesson, in consideration of $25,000 and other valuable assets, sold, transferred and assigned to the Volcanic Repeating Arms Company certain patents. From later records it was shown that they received 2,800 shares of stock in the company and $40,000 for the patents. Smith & Wesson were paid separately for the machinery and parts of arms then at Norwich. The $40,000 cash was to be paid in four installments.[2]

Oliver Fisher Winchester, 1810–1880

In their book, *The History of the Winchester Repeating Arms Company*, Veader and Earle relate a detailed history of the early finances and development of the company. They mention that, "In our files there is an instrument, dated July 26, 1856, from Courtlandt Palmer, Horace Smith, Daniel B. Wesson and the limited co-partnership of Smith & Wesson to the Volcanic Repeating Arms Company, certifying to the completion of the conditions expressed in the original contract of sale dated July 10, 1855, passing to the Volcanic Repeating Arms Company the absolute unconditional title to the patents referred to in the original document. Full payment for the machinery and stock at Norwich was made to Smith & Wesson."[3]

In the agreement of sale it was specified that, in addition to the transfer of the patents, there was transferred "also all and singular the interest of each and every of said parties of the first part in each and every of said patents and in all the rights, privileges and advantages secured thereby, and all renewals and extensions of said patents and of each and every one of them. And also all patents and patent rights which the said parties of the first part or either of them or their representatives or assigns, may or can obtain or acquire for Inventions or Improvements in firearms and ammunition or upon the matters already patented as aforesaid; already devised or hereafter to be devised by the said parties of the first part or any of them."[4]

All of this dry, rather boring legal manipulation and verbiage is of little interest to today's collector, but is nevertheless a part of the history of these companies. This latter clause in the sales contract was the cause of the litigation which later took place between the Winchester Repeating Arms Company and Smith & Wesson.

During August of 1855, all tools, fixtures and machinery and the approximately 500 to 600 finished and unfinished small-frame Smith & Wesson pistols were moved to a new location in New Haven. The first six months were spent in the assembly and marketing of these pistols and in the construction, purchase and modification of tooling and machinery for the new brass-frame pistols to be produced.[5] Experiments during that time indicated that a brass or bronze receiver would be easier to manufacture than the iron frame used in the Smith & Wesson arms.[6]

The Volcanic Repeating Arms Company was relatively large, employing some 50 full-time workers. The development of machinery since the days of Robbins & Lawrence made the pistol parts rather interchangeable, but filing and hand fitting were still needed for a more precise fit just as it was needed by the former Smith & Wesson Company. Subcontractors were used for the casting of the brass frames, pre-drilled barrels and rough-formed grips.[7] Some of the work done for the company was performed by the Whitney Arms Company at Whitneyville, which was really the first United States firm to develop a system for the mass production of gun parts.

Daniel Wesson continued as plant manager of the factory while it was at Norwich. His name appeared on the regular payroll until February 11, 1856, when he resigned. William C. Hicks, a Colt employee, was hired by Winchester to replace him.[8]

Some rather misleading reports of the company's progress were published in New Haven newspapers of the period. There were claims of astounding range and unerring accuracy and it was said that orders were crowding in upon the company from many quarters. The truth was the public was little more enthusiastic about the products of the Volcanic Repeating Arms Company than they had been for the Hunt, Jennings, Smith-Jennings or Smith & Wesson repeating firearms. Although there were some points in favor of these firearms, such as rapidity of fire, self-contained, generally waterproof ammunition and a sturdy, well-designed action, they were not enough to prevent a pile-up of unsold arms and mounting debts.[9]

Of the original 6,000 shares of stock, 2,800 shares had been transferred to Palmer, Smith and Wesson. After other disbursements, the company was left with a working capital of only $10,000 forcing the President, Nelson Gaston, and the Vice President, Oliver Winchester, to secure loans from local banks. They themselves loaned the company approximately

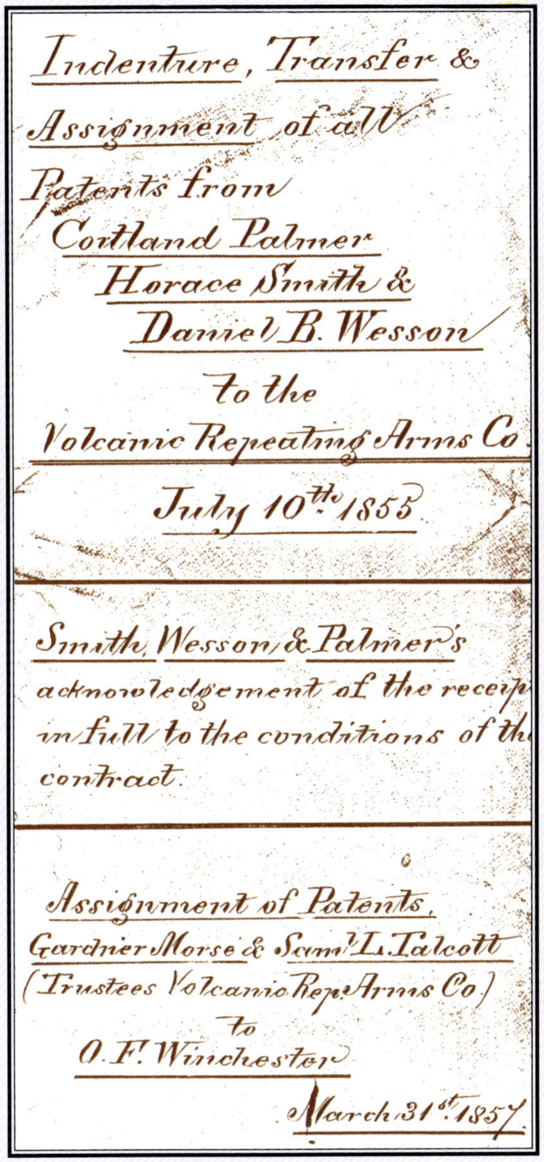
Indenture, Transfer &
Assignment of all
Patents from
Cortland Palmer
Horace Smith &
Daniel B. Wesson
to the
Volcanic Repeating Arms Co
July 10th 1855

Smith, Wesson & Palmer's
acknowledgement of the receip
in full to the conditions of th
contract.

Assignment of Patents,
Gardner Morse & Saml L. Talcott
(Trustees Volcanic Rep. Arms Co)
to
O.F. Winchester
March 31st 1857

Assignment of patents from Courtlandt Palmer, Horace Smith and Daniel B. Wesson to O.F. Winchester and the Volcanic Repeating Arms Company.

Herbert G. Houze collection

$20,000, secured by a mortgage. The firm of Winchester & Davies loaned an additional $1,500. J.W. Post, one of the original subscribers attempted to recover some $11,000 in loans he had made to the company. The Volcanic Repeating Arms Company could not cover these and other debts, and the Tradesman's Bank filed a petition to have the company declared bankrupt. Herbert Houze has made the observation that, while the collapse of the Volcanic Repeating Arms Company has been generally attributed to poor financial management, aggravated by poor sales, the root cause was its over capitalization in machinery during the concern's set-up in 1855.[10]

Oliver Winchester never lost hope for the future of a magazine repeater and refused to allow the total destruction of the company by the court. Nelson Gaston died in December of 1856, and Oliver Winchester became President of the company. Following the declaration of insolvency, he purchased the claims of Gaston's heirs and those of the Tradesman's Bank for only $40,000, which would be recognized today as a bargain. On March 15, 1857, the court awarded all assets of the Volcanic Repeating Arms Company to Oliver Winchester, including the assignments of all patents and the agreement with Smith & Wesson for rights to produce any future ammunition and firearms improvements they might invent or develop. The other stockholders received nothing for their investments.[11] Oliver Winchester's superior sales ability was tested when he gathered eleven subscribers, seven of whom had been shareholders in the Volcanic Repeating Arms Company, to capitalize the New Haven Arms Company, discussed in Chapter 5.[12]

GENERAL FEATURES OF THE PISTOLS

The Volcanic Repeating Arms Company abandoned the iron frames used in the production of Smith & Wesson's pistols for polished brass frames. This was the major difference between them. Brass frames, or more accurately, gunmetal, were found to be less expensive to manufacture and finish. They were made in three barrel lengths. Various barrel measurements have been given depending on the manner in which the measurement is made but are almost universally termed 8-inch, 6-inch and 16-inch. The latter was for a pistol-carbine with a detachable shoulder stock, a new addition to the product line. The company did not produce a small-frame pocket pistol of its own, likely because they inherited finished pistols and parts for small-frame pistols from Smith & Wesson. The barrels were octagonal rather than the part-octagon, part-round shape of the Smith & Wesson pistols. This was also a cost-saving measure since less lathe work was needed to produce the barrels. The bottom of the magazine tube remained open, as on the Smith & Wesson arms, to allow the magazine follower to function. This design flaw, allowing dirt and other debris to enter the magazine, was not to be corrected until the appearance of the Winchester Model 1866. Both the barrel and magazine were blued. Hammers and levers were generally case-hardened but some were blued. Spurs are never found on the levers of these pistols. Mechanically, the Volcanic Repeating Arms Company pistols operated in the same fashion as the Smith & Wesson firearms, with the internal parts being almost identical.

During the first six months of the company's existence, they produced no brass-frame pistols at all, but spent production time in assembling the 500 to 600 small-frame Smith & Wesson pistols from the parts acquired by the new company. The enterprise was shortlived, and during the last thirteen months of its existence, the Volcanic Repeating Arms Company had a labor force of about 50 workers, including seven women, and produced nearly 3,000 Navy pistols. The company supposedly did not produce carbines, but this interesting topic will be addressed later in the chapter.

The Volcanic Repeating Arms Company used a number of outside contractors for the production of the components of the pistols. The concept of using outside contractors was later used to great advantage by Oliver Winchester during production of firearms by both the New Haven Arms Company and, finally, the Winchester Repeating Arms Company. All firearms produced by the Volcanic Repeating Arms Company bear the February 14, 1854, patent date.

THE VOLCANIC NAVY PISTOLS

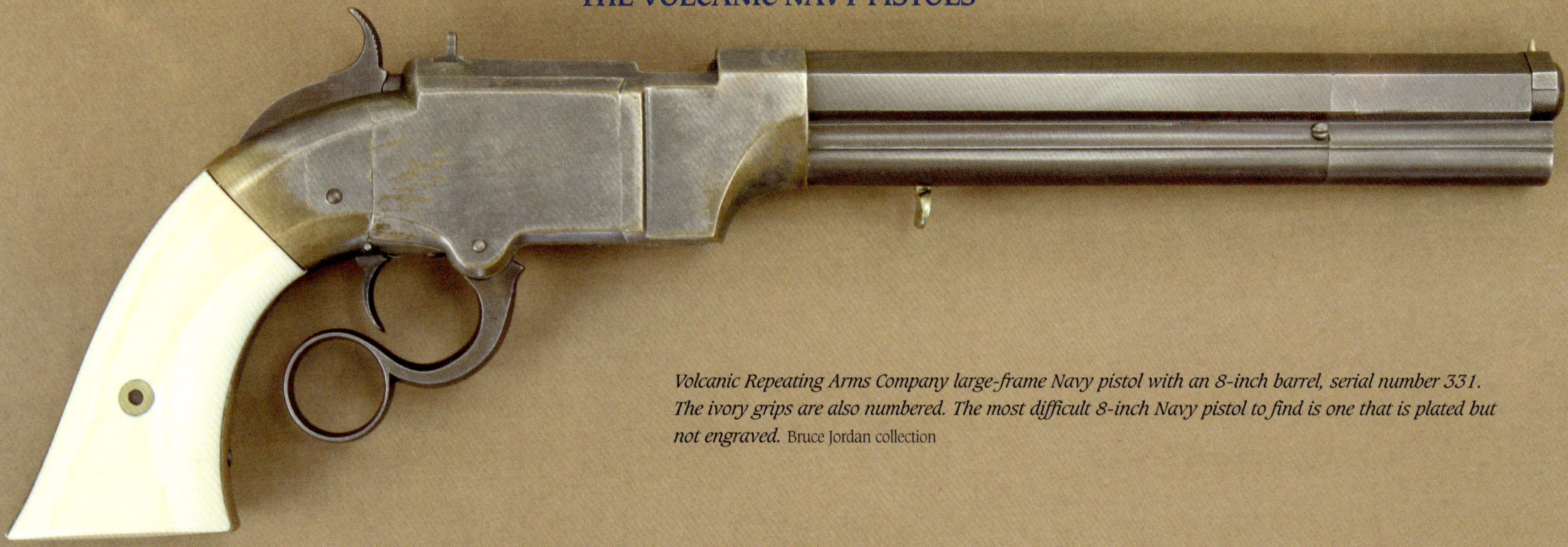

Volcanic Repeating Arms Company large-frame Navy pistol with an 8-inch barrel, serial number 331. The ivory grips are also numbered. The most difficult 8-inch Navy pistol to find is one that is plated but not engraved. Bruce Jordan collection

The large-frame Navy pistol was produced in two barrel lengths, an 8-inch "holster" model and a 6-inch "belt" pistol. They were virtually the same except for the barrel length and magazine capacity. The 8-inch pistol held ten rounds of the No. 2 Smith & Wesson Volcanic cartridges, and the 6-inch pistol held eight.

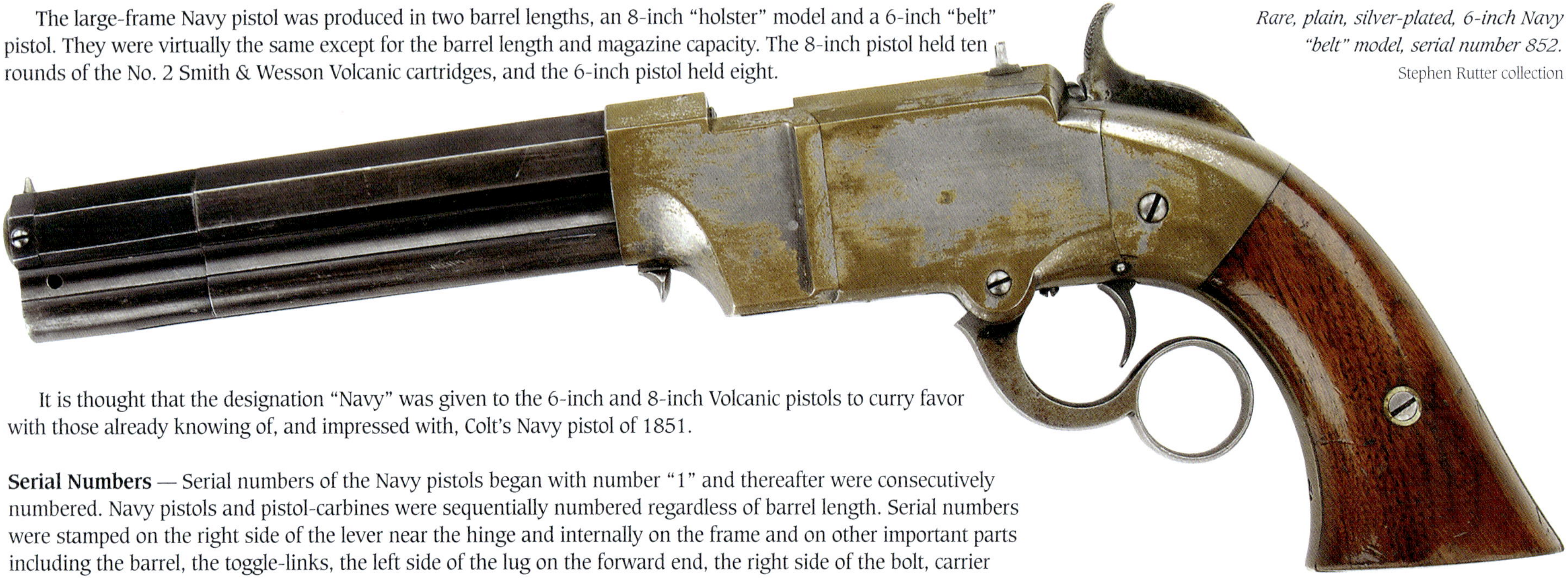

Rare, plain, silver-plated, 6-inch Navy "belt" model, serial number 852. Stephen Rutter collection

It is thought that the designation "Navy" was given to the 6-inch and 8-inch Volcanic pistols to curry favor with those already knowing of, and impressed with, Colt's Navy pistol of 1851.

Serial Numbers — Serial numbers of the Navy pistols began with number "1" and thereafter were consecutively numbered. Navy pistols and pistol-carbines were sequentially numbered regardless of barrel length. Serial numbers were stamped on the right side of the lever near the hinge and internally on the frame and on other important parts including the barrel, the toggle-links, the left side of the lug on the forward end, the right side of the bolt, carrier

Right: *Serial number "1418" stamped on the frame and grip.* Collection of Alden S. Bradstock, III

The 8-inch Navy "holster" model, serial number 1673.
Edmund Lewis collection

The 6-inch Navy "belt" model, serial number 826.
Bruce Jordan collection

and the inside of the sideplates. The inside of the grips were usually numbered as well. The serial numbers also functioned as assembly numbers, although some guns also had separate assembly numbers. Letters, such as "H", "L", "V", "X", and "Y", are found stamped on the left side of the lever on some pistols. They are of unknown meaning but perhaps represent inspector markings.

Volcanic Navy pistol with serial number "1669" stamped on toggle-link, sideplates and grip. Collection of Alden S. Bradstock, III

"Y" mark on lever.

Variations of engraving were common. Pictured here is the only known Volcanic Repeating Arms Company carbine, serial number 2683, with typical scrolls. Bruce Jordan collection

Below: *More elaborate engraving with a punched-dot background on a Navy pistol with ivory grips, serial number 1638.* Gerald Klaz collection

Engraving — Most Volcanic Repeating Arms Company pistols were engraved on the top and sides of the frame and backstrap with a pattern of leaves and scrolls. The engraving was done by several individuals who worked for the company and who did the engraving as an adjunct to their regular jobs of assembling and finishing. The engraving was simple and of moderate coverage, and while there are common features, there is some variation as one might expect. A few were not engraved, and fewer still had the frames silver plated. Only rarely is an engraved presentation firearm found. Engraved inscriptions are quite rare.

Barrels and Barrel Markings — The barrels of almost all pistols, pistol-carbines and the only known Volcanic Repeating Arms Company 25-inch carbine have a full rib between the barrel and the magazine. There is a partial rib on some 8-inch Navy pistols. This detail will be discussed further in Chapter Five. Volcanic Navy pistols were marked on the top flat:

THE VOLCANIC
REPEATING ARMS CO.
PATENT NEW HAVEN CONN. FEB.14.1854.

Five different dies were used to apply the barrel markings. Line placement and spacing of the stampings vary. The markings often varied in depth or were lightly impressed. On some pistols "VOLCANIC" or other words may be only partially legible or are completely worn away.

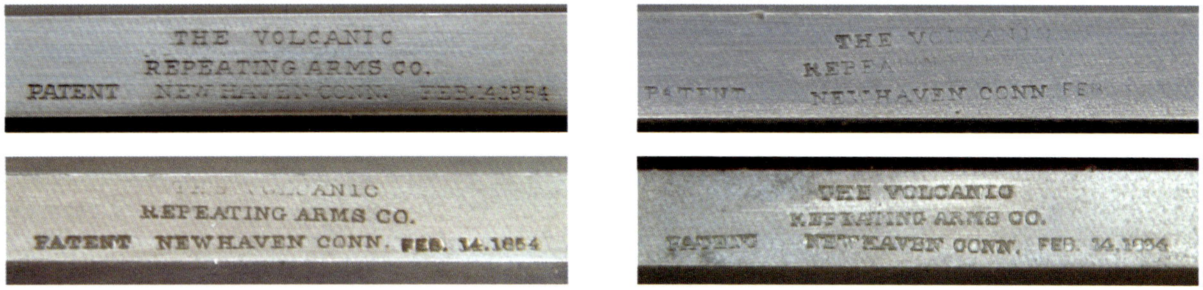

These four barrel markings show variation in depth, placement and range. They vary from full or partial strikes to some which are almost illegible.

Extractors — The extractors described in the earlier Smith & Wesson pistols were continued in the pistols of the Volcanic Repeating Arms Company with some variation. An improvement in the extractor was developed in 1856 by William C. Hicks, a former Colt employee, who was hired as plant manager to replace Daniel Wesson when he resigned on February 11, 1856. On March 10, 1857, Hicks was issued a patent for a "double nipple" for the firing and extraction of cartridges and he had this to say: *"...the end of the cartridge was closed with a metallic disc and cork containing fulminate between them. A blunt instrument projecting centrally from the breech pin was pressed through this cork and mashed the fulminate against the steel disc when it was fired by a blow from the hammer. The ball never extracted if there was a misfire, you had to push it out with a ramrod The steel disc after firing went out the muzzle sometimes and sometimes came back and caused trouble."*[13]

Above, left to right: *Single hooked extractor; Hick's "double nipple" with wear; erosion of single- and double-hook extractors by fulminate primer.* Donovan Williamson collection

The Hick's improvement proved less than ideal, which contributed to the Volcanic Repeating Arms Company's lack of success in sales.

Ejection Port Variations — Some Navy pistols have ejection ports with a shallow semicircular cut-out at the front and rear of the port, while others have a square cut-out at the front and a semicircular cutout at the rear. These cut-outs likely facilitated extraction of the cork base of the cartridge.

Above left is serial number 1050; on the right is serial number 1057. Note the change in ejection port shapes between these two close serial numbers. Donovan Williamson collection

Sights — The rear sight had a simple notch and was dovetailed into the receiver. The front site was similar to those on the Smith & Wesson pistols and usually was a thin silver cone or, in some cases, a small blade. They screwed into the top of the muzzle collar.

Magazine Blocks and Collars — The end of the barrel was crowned and extended beyond the magazine lock. A T-bar muzzle catch was used and had the same appearance as on the earlier Smith & Wesson pistols. Magazine block and collar dimensions varied as the barrel length increased to accommodate the longer magazine springs.

Gun	Length of Magazine block	Length of collar
6-inch Navy pistol	1.625 in.	0.250 in.
8-inch Navy pistol	2.125 in.	0.250 in.
16-inch pistol-carbine	2.625 in.	0.312 in.

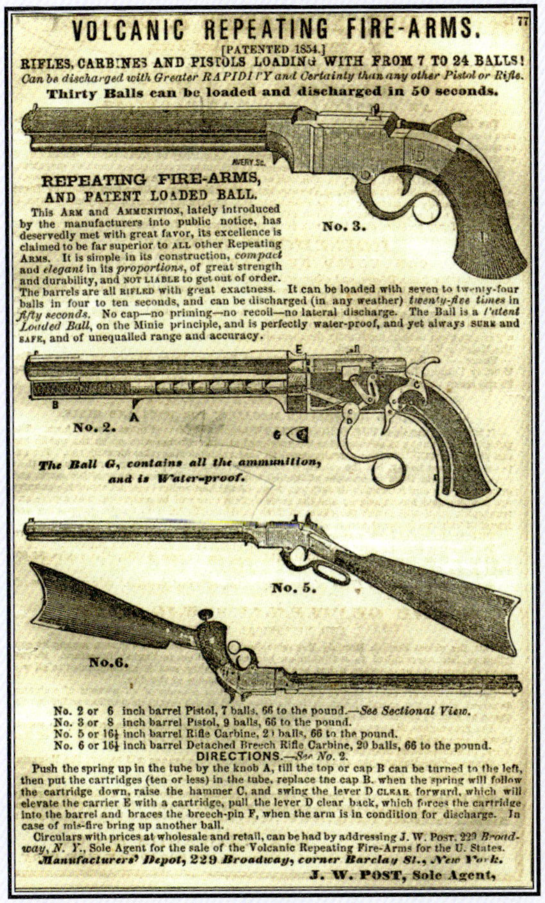

VOLCANIC REPEATING FIRE-ARMS.

[PATENTED 1854.]

RIFLES, CARBINES AND PISTOLS LOADING WITH FROM 7 TO 24 BALLS!

Can be discharged with Greater RAPIDITY and Certainty than any other Pistol or Rifle.

Thirty Balls can be loaded and discharged in 50 seconds.

REPEATING FIRE-ARMS, AND PATENT LOADED BALL.

This ARM and AMMUNITION, lately introduced by the manufacturers into public notice, has deservedly met with great favor, its excellence is claimed to be far superior to ALL other Repeating ARMS. It is simple in its construction, *compact* and *elegant* in its *proportions*, of great strength and durability, and NOT LIABLE to get out of order. The barrels are all RIFLED with great exactness. It can be loaded with seven to twenty-four balls in four to ten seconds, and can be discharged (in any weather) *twenty-five times in fifty seconds.* No cap—no priming—no recoil—no lateral discharge. The Ball is a *Patent Loaded Ball*, on the Minie principle, and is perfectly water-proof, and yet always SURE and SAFE, and of unequalled range and accuracy.

No. 2 or 6 inch barrel Pistol, 7 balls, 66 to the pound.—*See Sectional View.*
No. 3 or 8 inch barrel Pistol, 9 balls, 66 to the pound.
No. 5 or 16½ inch barrel Rifle Carbine, 2[illegible] balls, 66 to the pound.
No. 6 or 16½ inch barrel Detached Breech Rifle Carbine, 20 balls, 66 to the pound.

DIRECTIONS.—*See No. 2.*

Push the spring up in the tube by the knob A, till the top or cap B can be turned to the left, then put the cartridges (ten or less) in the tube, replace the cap B, when the spring will follow the cartridge down, raise the hammer C, and swing the lever D CLEAR forward, which will elevate the carrier E with a cartridge, pull the lever D clear back, which forces the cartridge into the barrel and braces the breech-pin F, when the arm is in condition for discharge. In case of mis-fire bring up another ball.

Circulars with prices at wholesale and retail, can be had by addressing J. W. POST, 229 *Broadway, N. Y.*, Sole Agent for the sale of the Volcanic Repeating Fire-Arms for the U. States.

Manufacturers' Depot, 229 Broadway, corner Barclay St., New York.

J. W. POST, Sole Agent,

"Circulars with prices at wholesale and retail, can be had by addressing J.W. Post, 229 Broadway, N.Y., Sole Agent for the sale of the Volcanic Repeating Fire-Arms for the U. States. Manufactures Depot, 229 Broadway, corner Barclay St., New York J.W. Post, Sole Agent."
Edmund Lewis collection

THE VOLCANIC PISTOL-CARBINE

A circular published in *The Tribune Almanac*, New York, 1857, illustrates the Volcanic pistol-carbine for the first time but does not price it. Note that it is described as a "Detached Breech Rifle Carbine."

In 1857, the depot in New York City was the primary distribution center for the Volcanic Repeating Arms Company's firearms, with J.W. Post listed as the company agent.

Taking a lead from the Colt's Manufacturing Company, shoulder stocks were made available for large-frame pistols with 16-inch barrels as well as Navy pistols with 8-inch barrels to transform them into pistol-carbines. It is estimated that only 300 or less of these were manufactured in both barrel lengths with the 8-inch pistol-carbine being very rare.

Volcanic pistol-carbine showing attachment of the stock to the grip, serial number 429. Edmund Lewis collection

The Volcanic Repeating Arms Company 16-inch pistol-carbine, serial number 429. Edmund Lewis collection

Left: *Adjustable rear sight of the pistol-carbine.* Edmund Lewis collection

This rare plated pistol-carbine has a wire stock that was added at a later date. Although the butt screw is very similar to those found on other Volcanic pistol-carbines, the back-strap lacks the hole for the upper pin of the normal shoulder stock. This stock has been with the firearm for many years and was most likely a period replacement as is the replaced rear sight. Alden S. Bradstock III collection

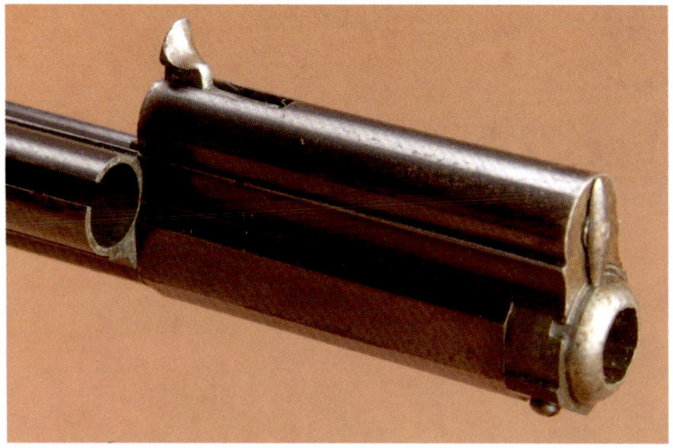

Muzzle and magazine block of a pistol-carbine, closed (at left) and open (at right). Edmund Lewis collection

These guns have a polished brass frame and buttplate and were seldom engraved. The barrel and magazine tube were blued as on the smaller pistols. Thc hammer and lever were case-hardened.

The 16-inch pistol-carbines and the rare 8-inch pistol-carbines had an adjustable rear sight. It was made of spring steel and dovetailed into the frame just to the rear of the ejection port. The rounded head of the inverted adjusting screw was knurled to facilitate adjustment, is threaded through the rear portion of the body of the sight and is seated in a recess at the top rear of the frame. Clockwise rotation of the adjusting screw raised the sight.

The magazine capacity was twenty .41 caliber No. 2 Smith & Wesson Volcanic cartridges; eighteen in the magazine and one each in the carrier and the chamber. The pistol-carbine's serial number is found in the same locations as on the Navy pistols, and both the buttplate and yoke of the shoulder stock were also numbered. Pistol-carbines are often found with shoulder stock serial numbers that do not match.

Volcanic Pistol-Carbine Variations — In his book *Evolution of the Winchester*, R. Bruce McDowell illustrates several variations produced by the Volcanic Repeating Arms Company that are rarely, if ever, seen by today's collector. Included are an 8-inch pistol-carbine with a detachable factory stock and another 8-inch pistol-carbine with a non-factory one-piece stock made of black walnut. It has the regular rear sight of an 8-inch Navy pistol rather than an adjustable sight, which indicates that it was converted from an 8-inch Navy pistol. These firearms are in the Nunnemacher Collection of the Milwaukee Public Museum and are not currently on display.

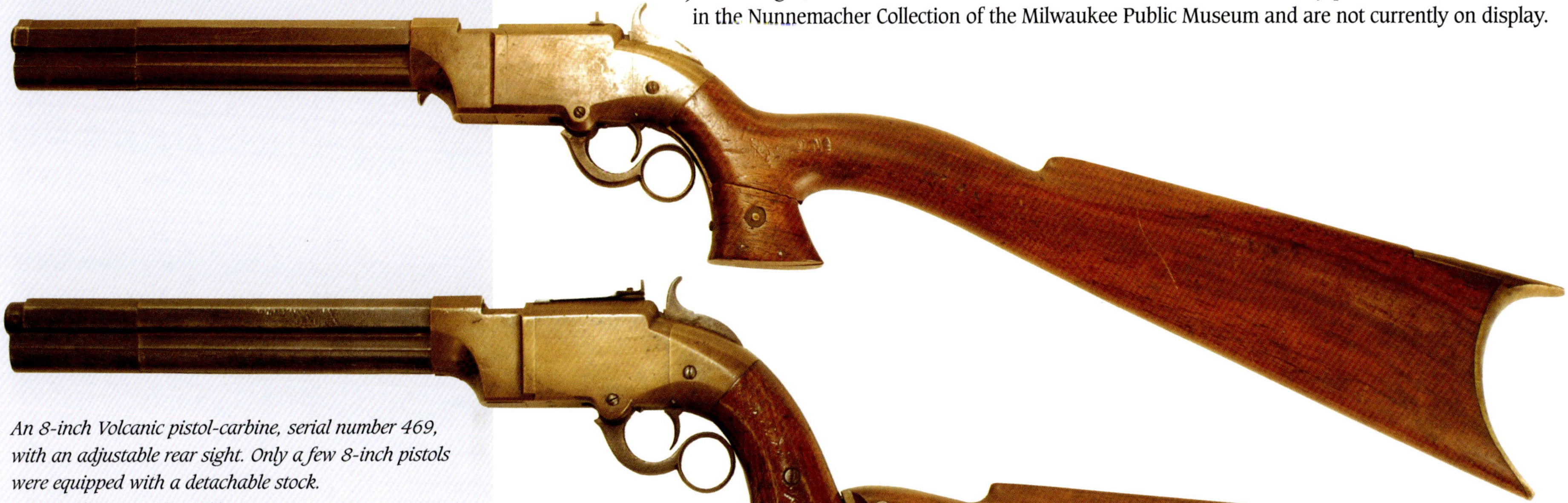

An 8-inch Navy pistol converted to a pistol-carbine, serial number 263. The unusually long one-piece black walnut stock and buttplate is non-factory. It has an unusually long fixed rear sight. Courtesy of the Milwaukee Public Museum, Nunnemacher Collection

An 8-inch Volcanic pistol-carbine, serial number 469, with an adjustable rear sight. Only a few 8-inch pistols were equipped with a detachable stock. Courtesy of the Milwaukee Public Museum, Nunnemacher Collection

THE VOLCANIC CARBINE

In his *Evolution of the Winchester*, published in 1985, R. Bruce McDowell states that he had never seen a carbine with a Volcanic-marked barrel. He tells us that, "the Volcanic Repeating Arms Company existed for only nineteen months, from July 1855 to February 1857, and during the first six months, the remaining 500 or 600 Smith & Wesson small-frame pistols were assembled and finished, and no brass-frame pistols or carbines were produced. During the remaining thirteen months, some fifty employees and a handful of outside contractors produced approximately 3,000 Navy pistols leaving little time or manpower for the tooling and production of carbines. There is no question that the management and marketing people considered adding carbines to their products, as shown by an 1856 advertisement and letterhead. These may, however, be nothing more than an effort to test the market for acceptance and popularity." He continues, "In all research on the subject, it has not been possible to find a single carbine or rifle with Volcanic barrel markings" and that "none were manufactured by the Volcanic Repeating Arms Company as standard production models and that if any were produced they were made late in 1856 on a special order or experimental basis."[14]

An 1856 advertisement listed Volcanic carbines in three nominal barrel lengths: 16, 20 and 24 inches, with capacities of 20, 25 and 30 cartridges. Also, the letterhead of J.W. Post, 23 John Street, New York, who in August 1856, claimed to be the sole agent for the Volcanic Repeating Arms Company, stated "VOLCANIC REPEATING RIFLES, CARBINES AND PISTOLS WILL FIRE 30 SHOTS PER MINUTE. AMMUNITION WATERPROOF."

These two references would seem to indicate that Volcanic rifles or carbines were produced by the Volcanic Repeating Arms Company. "VOLCANIC"-marked rifles and carbines would probably have 16-inch, 21-inch and 25-inch barrels if similar types made by the later New Haven Arms Company are an indication.[15] A carbine produced by the Volcanic Repeating Arms Company with a "VOLCANIC" barrel marking is an extreme rarity, and after the assembly of the remaining small-frame Smith & Wesson

An extraordinarily rare 25-inch Volcanic carbine, serial number 2683. Bruce Jordan collection

pistols, the short time remaining in the life of the company (about thirteen months) left little time or manpower for the production of long arms.

That said, a "VOLCANIC"-marked carbine is shown on page 62. Serial number 2683 has a 25-inch, .41 caliber barrel with a brass frame and buttplate that are engraved and silver plated with a walnut stock. The configuration of the barrel and magazine tube are typical of Volcanics with a Type I ribbed barrel. The barrel has a rib between the barrel itself and the magazine. The barrel marking is:

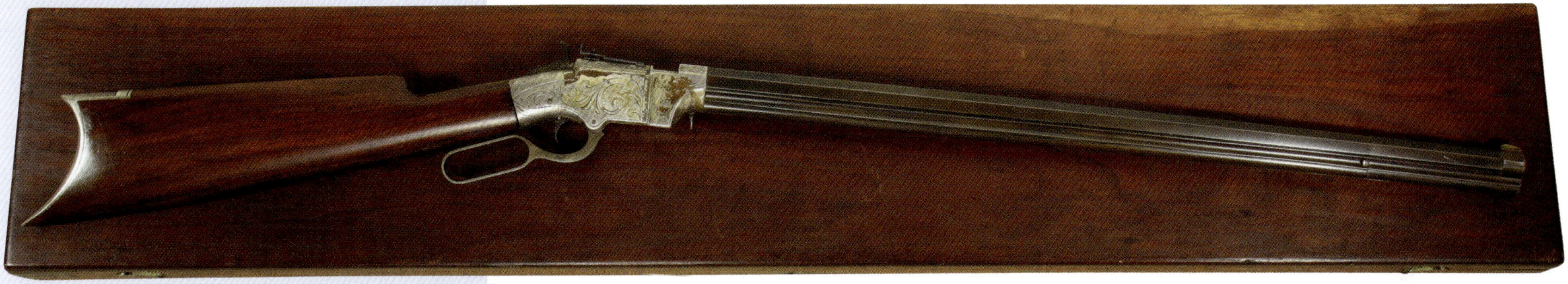

Full-length view of rare 25-inch Volcanic carbine, serial number 2683, with Type I full-ribbed barrel.
Bruce Jordan collection

THE VOLCANIC
REPEATING ARMS CO.
PATENT NEW HAVEN CONN FEB. 14, 1854

The serial number is inverted on the right side of the lower tang and is also found on the sideplates, lever, rear of the magazine block, buttstock and buttplate. An assembly number "149" is also present on the left tang.

McDowell further tells us that all firearms produced by the Volcanic Repeating Arms Company have ribbed barrels, while all the carbines, pistol carbines and small-frame pistols produced by the New Haven Arms Company have barrels without ribs, and further, "Had the Volcanic company produced any carbines, the barrels of those carbines would have ribs."[16]

Only the configuration of the carbine barrels, ribbed or no rib, will establish their origin, regardless of barrel markings. Serial numbers are meaningless in establishing the manufacturer of parts, since parts made by the Volcanic Repeating Arms Company without serial numbers were later assembled and numbered by the New Haven Arms Company with both Volcanic and New Haven frames.[17] Carbines have been fraudulently made up using a 16-inch barrel from a Volcanic Repeating Arms Company pistol-carbine.

Barrel with full rib.

Half-ribbed barrel.

Non-ribbed barrel.

Far right: *The larger front sight on the New Haven Arms Company carbine at the top. Note the shorter magazine block and collar of the pistol-carbine is at the bottom.*
Collection of Edmund Lewis

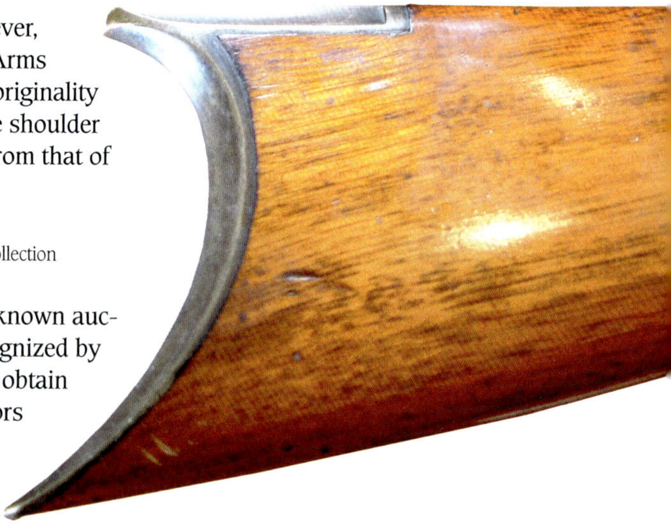

The Volcanic Repeating Arms Company pistol-carbine, however, had a shorter magazine block than that of the later New Haven Arms Company carbine. This is an important feature in determining the originality of a carbine produced by the Volcanic Repeating Arms Company. The shoulder stock was detachable and the shape of the buttplate differed slightly from that of the New Haven Arms Company.

Carbine buttplate at left, Pistol-carbine buttplate at right. Edmund Lewis collection

The 25-inch carbine shown, although properly described by a well-known auction house with the designated "VOLCANIC" barrel marking, was recognized by only a few knowledgeable collectors, one of whom was fortunate to obtain the gun. May such good fortune sometime find us all. The authors know of no other Volcanic Repeating Arms Company carbine with either a 16-inch, 21-inch or 25-inch barrel.

A Volcanic Navy pistol, serial number 884, with a 6-inch barrel. The case contains .41 caliber cartridges in a tin cartridge box. Courtesy of James D. Julia, Inc.

Smith & Wesson First Model, First Issue .22 short RF revolver, serial number 1225.
Courtesy of Greg Martin Auctions

Production Numbers — Production of the Navy 8-inch "holster" pistol is estimated to be 1,500, and production of the small Navy "belt" pistol is estimated to be 1,200. The total production of the pistol-carbines in both the 16-inch and 8-inch models is estimated to be less than 300. Though the Volcanic Repeating Arms Company's pistol had a reputation for misfiring and the cartridge was woefully weak, many were still sold, as it was the first cartridge repeating firearm available.[18]

Cases — The Volcanic Repeating Arms Company did not produce cases themselves but packaged and shipped their pistols in heavy black cardboard boxes. These boxes were most likely made by the Kingsbury Box Company of South Coventry, Connecticut. Presumably, the few carbines produced were shipped in similar boxes. None, however, are known. Some Volcanic firearms made either on special order or for presentation were cased in walnut or mahogany. A few had a silver presentation plaque on the top. These cases were not manufactured by the company but were produced by outside contractors and built to the company's specifications. They are partitioned in the English style and often contained one or two boxes of No.1 or No. 2 PISTOL CARTRIDGES. The cases were lined with fabric in a variety of colors, including green, burgundy, red or blue. A few cases had a narrow space for a rod at the front or back. The rods were not called "cleaning rods" by the company, but appear to have been primarily for clearing the action in the event of a misfire. As mentioned previously in Chapter Three, the ammunition labels state:

"If, by accident, a ball should miss fire, it must be carefully pushed back with the rod, leaving the carrier down."

The rods for pistols were usually of dark rosewood and were sized to fit the barrel length. Longer rods for the carbines were of hickory or other hardwoods. One end had a brass ferrule so the rod was also meant for cleaning the bore. Wood cases were made after the firearm left the factory, and any period case is very desirable. The number of cardboard boxes that have survived are far less than the wooden cases. More examples of these boxed and cased firearms will be illustrated in Chapter Six.

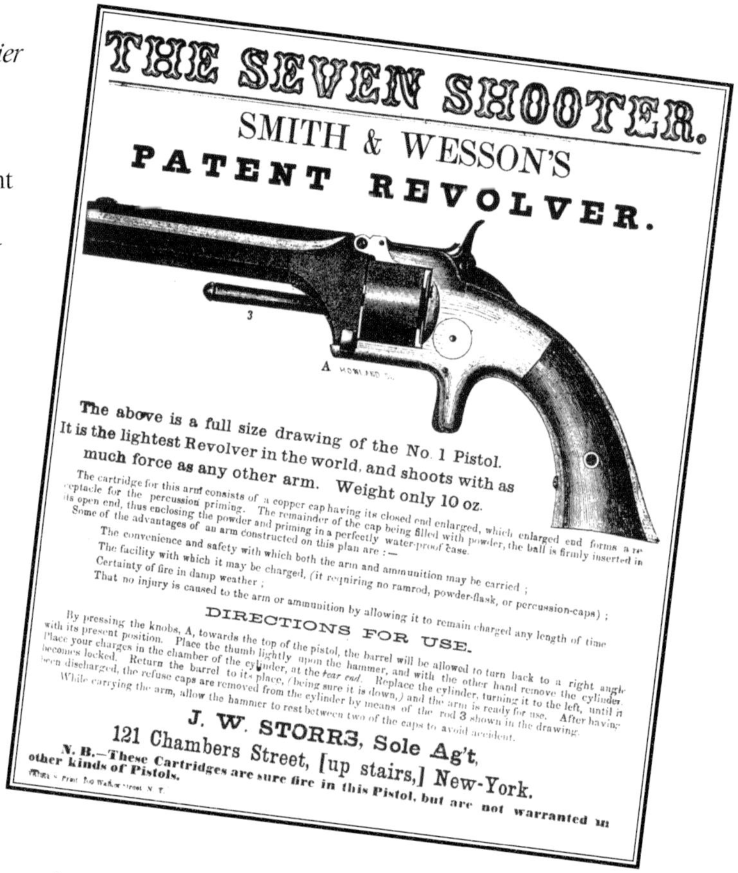

"The Seven Shooter."
From Jinks, *History of Smith & Wesson*

NEW BEGINNINGS FOR SMITH & WESSON: THEIR SECOND PARTNERSHIP

The subsequent failure of the Volcanic Repeating Arms Company resulted in two historically divergent and important pathways. Oliver Winchester, with the capable assistance of B. Tyler Henry, would go on to form the New Haven Arms Company, while Horace Smith and Daniel Wesson would later form their second partnership to produce the first of their revolver pistols.

Smith must have become disenchanted with the Volcanic Repeating Arms Company since he left the firm and returned to Springfield, Massachusetts, to work with his brother-in law. Wesson remained with the company and continued to experiment with his idea for a Flobert-style metallic cartridge and various types of arms, including a wooden model revolver which was designed to fire a .22 rimfire-style cartridge.[19] Fortuitously, the Rollin White patent for a bored-through cylinder had been issued, and needing this essential feature for his model pistol, Wesson contacted White on October 31, 1856, inquiring whether they could reach an arrangement to use his patent. A meeting took place on November 17, 1856, at which D.B. Wesson and Horace Smith signed an agreement with White for an exclusive license to manufacture a revolver with a bored-through cylinder.[20]

After this agreement, in which White would not have the right to manufacture firearms under his patent and would defend it against all infringers, Smith and Wesson entered their second partnership and formed the Smith & Wesson Revolver Factory of Springfield, Massachusetts. They introduced a 7-shot .22 rimfire revolver in 1857, which was manufactured in various forms until 1881. This small revolver marked the beginning of the era of the self-contained cartridge. Roy Jinks summarized this era of the partners as follows:

"When Smith & Wesson abandoned lever-action magazine arms to go onto distinguished careers as makers of revolving cylinder pistols, they left behind something of greater value than they may have realized — the prototype of a gun which helped build one of the greatest names in gun making history."

chapter five

VOLCANIC

REPEATING FIRE-ARMS.

PATENTED 1854.

The New Haven Arms Company

Have recently obtained the entire control of the above Arms, and are now manufacturing them in the most perfect manner, and are prepared to furnish the trade on favorable terms.

The rapidity of execution of this Arm places it beyond all competition. The thirty shooter can be loaded and fired in less than one minute—a quickness and force of execution which is as much superior to the best revolvers as they are to the old muzzle loading single shooters.

The Ammunition is water-proof, hence it can be used in any weather, or loaded and hung up for months, or laid under water, and then fired with certainty.

Its safety from accidental discharge is a great consideration in its favor; for while the magazine (a tube running the whole length of the barrel) may be filled with balls, and thus the gun, in fact, be loaded from breech to muzzle, it is yet impossible, from any carelessness in handling, to discharge it. *Its construction* is simple and its workmanship most perfect, hence it is not easily got out of repair.

Its proportions are light, elegant, and compact, and the barrels are all rifled with great exactness. It requires no cap nor priming, no bullet mould nor powder flask. The powder and cap is contained in a loaded "minnie" ball of the best form and proportions, and is as sure as the best percussion caps.

It shoots with accuracy and greater force than any other arm can with double the powder used in this. Directions for use accompany each Arm. Balls are packed in tin cases, 200 each.

After this date, the prices will be as follows, viz.:

No. 1, 4 inch Pocket Pistol,	$12 00,	Plated and Engraved, $13 50,	Carrying 6 Balls.
No. 1, 6 " for Target Practice,	13 50,	" " " 15 00,	" 10 "
No. 2, 6 " Navy Pistol,	18 00,	" " " 20 00,	" 8 "
No. 2, 8 " " "	18 00,	" " " 20 00,	" 10 "
No. 2, 16 " Carbine,	30 00,	" " " 33 00,	" 20 "
No. 2, 20 " "	35 00,	" " " 38 00,	" 25 "
No. 2, 24 " "	40 00,	" " " 43 00,	" 30 "

AMMUNITION.

No. 1 Balls, 130 to the Pound, $10 per Thousand. No. 2 Balls, 66 to the Pound, $12 per Thousand.
(No. 1 Arms, require No. 1 Balls. No. 2 Arms, require No. 2 Balls.)

A liberal discount to the trade. Your orders are respectfully solicited.

NEW HAVEN ARMS COMPANY,
NEW HAVEN, CONN.

Office and Depot for Sales, 267 Broadway, New York.
JOSEPH MERWIN, Agent.

NEW YORK, MAY 1, 1859.

The New Haven Arms Company's Volcanic Arms

Failure loomed on the horizon for Oliver F. Winchester and his fellow stockholders. The Volcanic Repeating Arms Company was in financial turmoil following the death of its president, Nelson Gaston in December of 1856. The company was unable to pay notes that came due and, finally, the company was declared insolvent in February of 1857. Oliver Winchester, however, was not a man to be dissuaded by failure and formed the New Haven Arms Company on May 1, 1857, with himself as president and treasurer. He was assigned by the court the major assets of the defunct Volcanic Repeating Arms Company. Oliver Winchester was enamored of the "VOLCANIC" name and retained it for marketing the firearms produced by the new company.

The manufacture of lever-action pistols was resumed, but some important changes were made, and by the end of 1857, small-frame pistols with two barrel lengths and carbines in three barrel lengths were added to the product line. Manufacturing was begun in the same New Haven location that had been rented by the Volcanic Repeating Arms Company. In June 1859, production facilities were moved to a larger building at No. 9 Artizan Street in New Haven, and by 1860, the company's workforce had grown to over seventy workers.[1]

The new company continued to use the same inefficient and unreliable self-contained cartridge and, not surprisingly, the problems with the ammunition continued to plague the company as it had also plagued both the Smith & Wesson and the Volcanic Repeating Arms Company. The New Haven Arms Company was quickly running out of working capital, and by 1861, the company was on the verge of bankruptcy.

The one bright star in the firmament was B. Tyler Henry who, in 1857 or 1858, was appointed plant manager because of his experience in the design of repeating firearms and, more importantly, with the development of ammunition. Henry realized that the lever-action mechanism was sound, but what was needed was a reliable and powerful self-contained metallic cartridge. Starting in 1858, he spent years in developing such a cartridge and modifying the Volcanic lever-action mechanism to accept it. His talent in this endeavor would soon be of untold benefit to the company as we shall later see. For the present, however, despite the company's other problems, Henry kept things running smoothly and kept the production lines moving.

With the company in the capable hands of B. Tyler Henry, Oliver Winchester now had the time and interest as a businessman to direct his own efforts towards advertising and building markets as well as attempting to establish dealers and sales agents in key locations such as New York City and Baltimore, Maryland. J.W. Storrs, J.W. Post and Joseph Merwin were the best-known agents and were listed on the company's circulars. R. Bruce McDowell tells us that discounts were offered ranging from 20 percent to 30 percent, depending on volume, with an additional 5 percent discount for cash. Ammunition carried a fixed dealer discount of 20 percent with minimum orders of $100.[2] Volcanic pistols had a reputation for misfiring, and the cartridge was woefully weak, but it had the advantage of being the first cartridge repeater available, and despite its drawbacks, many were sold.

Volcanic firearms were not only marketed in the United States but also throughout Europe and in Central and South America. Frank Leslie in his *Illustrated Newspaper* of October 9, 1858, wrote that, "The Volcanic arms, wherever exhibited, have attracted great attention, and have been extensively rewarded with premiums... A case of these arms is now exhibited at the Fair in the Crystal Palace, where their novel construction and capabilities attract numbers of inquiries. The arms are already extensively exported to South America, Cuba, California and Oregon, as well as to France and Spain."[3]

Prices of Arms and Ammunition and Distribution — Circulars and broadsides that were published by the company listed the choice of arms available, along with their price and that of the ammunition. The few special features then available were also noted. Other extra features such as ivory grips, special inscriptions and plating (other than silver) were not mentioned in the circulars. Also, wood cases were not listed, but all were available. The New Haven Arms Company circular of May 1, 1859, listed the prices for firearms and ammunition as shown below. Note that the pistol-carbines were not mentioned or priced on this circular:

No. 1,	4-inch Pocket Pistol	$12.00,	Plated and Engraved,	13.50,	Carrying	6	Balls.
No. 1,	6-inch for Target Practice	13.50,	" " "	5.00,	"	10	"
No. 2,	6-inch Navy Pistol	18.00,	" " "	20.00,	"	8	"
No. 2,	8-inch Navy Pistol	18.00,	" " "	20.00,	"	10	"
No. 2,	16-inch Carbine	30.00,	" " "	33.00,	"	20	"
No. 2,	20-inch Carbine	35.00,	" " "	38.00,	"	25	"
No. 2,	24-inch Carbine	40.00,	" " "	40.00,	"	30	"

AMMUNITION

No. 1 Balls, 130 to the Pound, $10 per Thousand — No. 2 Balls, 66 to the Pound, $12 per Thousand
(No. 1 Arms, require No.1 Balls. — No. 2 Arms, require No.2 Balls)

A liberal discount to the trade. Your orders are respectfully solicited.

NEW HAVEN ARMS COMPANY,
NEW HAVEN, CONN.

GENERAL FEATURES OF THE FIREARMS

The main differences between Volcanic Repeating Arms Company and New Haven Arms Company firearms were the barrel shape and marking, location of the serial numbers and different shape of the pistol butt.[4] The barrels were usually Type III, without ribs. The frames were polished brass, and the internal parts had no discernable difference in form or function. The barrels and magazines, carriers and triggers continued to be brightly blued with the hammer and lever being case-hardened. As with the previous Volcanic arms, there were no lever spurs. No mechanical improvements were made in the new company's guns. The product line was expanded with the addition of a 4-inch Pocket pistol similar to the one produced earlier by Smith & Wesson. A 6-inch Target pistol was also added along with carbines with 16½-inch, 21-inch and 25-inch barrels, although the company's advertising usually referred to them as 16, 20 and 24 inches. Only a few pistol-carbines were produced from barrels and stocks that were left over from The Volcanic Repeating Arms Company.

Serial Numbers — Serial numbers, as with the Volcanic Repeating Arms Company, began at number "1", and whether the firearm was a pistol, a carbine or a pistol-carbine, they ran in a consecutive sequence until the end of production. Pistols were stamped on the bottom of the left side of the frame under the grip. The carbines had serial numbers on the side of the lower tang, under the stock and on the inside of the buttplate at the top and on the wood. In addition, the screws attaching the buttplate and tangs to the stock were usually marked with the serial number on their side below the head. Serial numbers do not appear on the levers of the New Haven Arms Company's pistols or carbines. They may be found, however, on the levers of guns completed by the New Haven Arms Company from parts obtained from the earlier company.[5] The full serial number is usually found on the major internal parts as were found on pistols of the preceding company. Serial numbers, how-

The action closed (above) and open (below) on a 6-inch Target pistol. Edmund Lewis collection

ever, will not establish the manufacturer of parts since Volcanic parts without serial numbers were assembled and numbered by the New Haven Arms Company. Assembly numbers were not required since each firearm was separately hand-fitted and assembled. An unusual "X" mark is infrequently found stamped on the left side of the lever, below the screw. Its significance is unknown.[6] Other letters found on the left side of the lever also include "L", "V", "H" and "Y" and were perhaps inspector's stamps.

Barrel Markings — The term "VOLCANIC" was considered by Oliver F. Winchester to have great marketing appeal since it was prominently displayed in the title of the New Haven Arms Company advertising circulars and broadsides. However, "VOLCANIC" was deleted from the earlier barrel marking of the Volcanic Repeating Arms Company. The new marking then read either:

"NEW HAVEN CONN. PATENT FEB. 14, 1854"

or

"NEW HAVEN CONN.
PATENT FEB. 14, 1854"

The words "NEW" and "HAVEN" were often run together forming a single word, "NEWHAVEN". Occasionally, "CAST-STEEL" is found in the marking, applied with a separate stamp. Variation in the depth of the stamping is common.

Above: *The typical New Haven Arms Company barrel marking.*

Variations in barrel markings.

Ejection Ports — The Navy pistols have a rectangular shape with a shallow cut-out only at the rear of the ejection port. The Target pistols and small-frame Pocket pistols, however, may show a cut-out at both the front and rear of the port.

Grips — There is a slight but distinct difference between the grips of the pistols that were produced by the Volcanic Repeating Arms Company and those later manufactured by the New Haven Arms Company. Navy and Target pistols had less flaring at the bottom of the backstrap than on the pistols of the earlier Volcanic Repeating Arms Company.[7] They were usually of walnut, but some were rosewood. Rarely, pistols were equipped with ivory grips and very rarely they were carved. The grips usually had stamped serial numbers, but some were marked in pencil.

Sights — The standard front sight on the earlier Navy pistols with one-digit, two-digit or three-digit serial numbers is a short, flat blade. This was changed at approximately serial number 1000 to a simple silver cone which did not require milling and was easier to install.[8] The rear sight of the pistols was a simple notched sight set into a dovetail at the rear of the frame. The front sight of the carbines was usually a German silver blade inset into a groove that was machined at the top of the magazine collar. Some carbines had silver cone front sights. Carbine rear sights were adjustable and were the same as found on the earlier Volcanic pistol-carbines. There was little variation.

Levers — The pistol levers were either case-hardened or blued and were without spurs. Carbine levers had a much larger finger loop, which were also case-hardened or blued.

Muzzles and Magazine Blocks — Muzzles were crowned, and the magazine catch was a T-bar. The following magazine block dimensions for the pistol, pistol-carbine and 16-inch carbine were recorded by R. Bruce McDowell in his book *Evolution of the Winchester*:

The Volcanic Repeating Arms Company grip at the top has more flair at the butt than the New Haven Arms Company grip on the bottom.

Magazine Block Dimensions		
Gun Description	**Length of Magazine Block**	**Length of Collar**
4-inch Pocket Pistol	1.000 inches	0.187 inches
6-inch Navy Pistol	1.625 inches	0.250 inches
8-inch Navy Pistol	2.125 inches	0.250 inches
16-inch Pistol-Carbine	2.625 inches	0.312 inches
16-inch Carbine	4.500 inches	0.625 inches
Smith & Wesson Small-Frame	1.125 inches	0.187 inches
Smith & Wesson Large-Frame	1.500 inches	None Used

Carbine Stocks and Buttplates — The rear of the buttplates had a top flat and two angled flats. Standard buttplates were of brass and were engraved with a design in keeping with that on the frame, sideplates and upper tang. If the frame was plated and engraved, so was the buttplate. Stock and buttplate screws were oval, indexed and often, but not always, numbered to the gun. Checkered stocks are rarely found. Although they may have been checkered at the factory, this extra was not listed in the company's advertising, and it is more likely that checkering was done after the firearm left the factory. The checkering shown in the photo below is coarse, and the design is typical for the period. Dimensions of the carbine and pistol-carbine buttplates were slightly different.[9]

Very rare checkered stock on a 25-inch New Haven carbine, serial number 863.
Raymond B. Bentley collection

Typical engraving on the frame of this 16-inch carbine, serial number 3039. This is a late-production carbine.
Edmund Lewis collection

Production Volumes — Approximately 3,370 pistols and carbines were produced by the New Haven Arms Company. Of these, it is estimated that 850 small-frame Pocket pistols were produced. Approximately 1,000 large-frame Navy pistols were produced with 8-inch barrels and about 300 were produced with 6-inch barrels. About 1,000 carbines in all three barrel lengths were manufactured.[10] A 16-inch carbine, serial number 3046, is in the Cody Firearms Museum. The highest serial number known is 3062. Approximately 225 Target pistols with a 6-inch barrels were produced between 1857 and 1860.

THE NEW HAVEN NO. 1 POCKET PISTOL

As previously stated, the New Haven Arms Company added production of a small-frame, short-barreled pistol similar to one produced by Smith & Wesson, but dropped from the line by the Volcanic Repeating Arms Company. Like the earlier Smith & Wesson small-frame pistol, it was .31 caliber but had a slightly shorter 3½-inch octagonal barrel, although a nominal barrel length of 4 inches was given in the company's advertising. The frame was brass rather than steel, and the bottom of the grip was flat rather than rounded. The frame, cartridge follower, lever and sights were smaller than those found on the Navy pistol. There is little variation in these pistols. The magazine held six .31 caliber No. 1 cartridges. Barrels and magazines were blued rather than browned and a single screw secured the lever and sideplates rather than two as on the Smith & Wesson small frame. Approximately 850 were produced in the years from 1857 to 1862.[11]

Engraved sideplates and frames were much more common on the small-frame pistol than were plain, polished frames even though the extra cost for engraving and plating was $1.50. Plain frames are very uncommon.

A pair of New Haven Pocket pistols with engraved and plain frames. Most small-frame pistols were engraved and plated. Those with plain frames are rare. Top, serial number 731; bottom, serial number 292. Gerald Klaz collection

Right: *An exceptional silver-plated, engraved small-frame New Haven Pocket pistol, serial number 1268.* Bruce Jordan collection

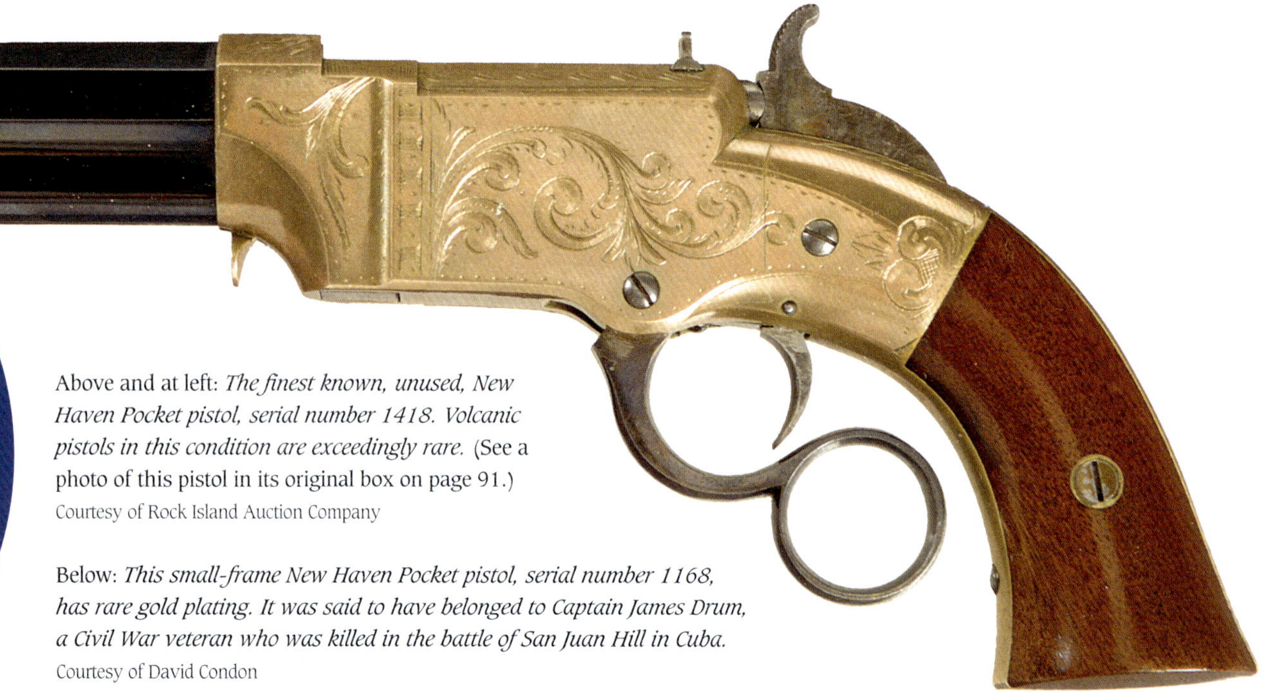

Above and at left: *The finest known, unused, New Haven Pocket pistol, serial number 1418. Volcanic pistols in this condition are exceedingly rare.* (See a photo of this pistol in its original box on page 91.)
Courtesy of Rock Island Auction Company

Below: *This small-frame New Haven Pocket pistol, serial number 1168, has rare gold plating. It was said to have belonged to Captain James Drum, a Civil War veteran who was killed in the battle of San Juan Hill in Cuba.*
Courtesy of David Condon

Right: *Silver-plated and engraved New Haven Pocket pistol with ivory grips and a different style of engraving with a blank riband.* Gerald Klaz collection

Below: *Two small-frame silver-plated Pocket pistols showing the variety of engraving patterns. Plating and engraving were an additional $1.50 over the $12.00 cost of the pistol.* Gerald Klaz collection

THE NEW HAVEN NO. 1 TARGET PISTOL

The .31 caliber 6-inch Target pistol is rare with only about 225 having been manufactured between 1857 to 1862.[12] It has a much smaller frame, grips, lever and cartridge follower as well as smaller sights than the Navy Pistol with a 6-inch or 8-inch barrel. Another difference was the lack of external retaining spring mounting screws on the bottom of the frame. Internal screws served the same purpose. Except for the barrel length, it was identical to the 3½-inch Pocket pistol. The barrel marking was uniform and easily read and contrasted with some other models, which sometimes were very lightly stamped.

The capacity was ten .31 caliber No.1 Pistol Cartridges; eight in the magazine and one each in the carrier and the chamber. Most of this model had a plain, polished brass frame. Target pistols with engraving are much more rare than those with plain frames.[13] The Target pistol cost $13.50; plating and engraving were an additional $1.50.

Edmund Lewis collection

Bruce Jordan collection

An engraved New Haven Target pistol, serial number 553, and one with a plain frame, serial number 491. Those with engraving are more rare than those without.

A collection in itself of four New Haven Target pistols. From top, serial numbers 803, 950, 1151 and 1953.
Donovan Williamson collection

A 6-inch New Haven Target pistol with plain frame, serial number 1053.
Ray Bentley collection

THE NEW HAVEN NO. 2 NAVY PISTOLS

The large-frame Navy pistols were produced with 8-inch and 6-inch barrels with the latter being more uncommon. Approximately 1,000 were manufactured with 8-inch barrels and only about 300 were made with 6-inch barrels.[14] Both were manufactured from 1857 to 1862. The finish of the frame, barrel assembly, trigger, hammer and lever was the same as on the small-frame pistols. The caliber was .41, and the capacity of the 8-inch pistol was ten No. 2 cartridges, and the 6-inch model held eight No. 2 cartridges. Navy pistols were usually not engraved or plated. The reason for this is unknown. One might surmise that the added cost of plating and engraving discouraged using these finishes, but note that most Pocket pistols with 3½-inch barrels were engraved.

Most New Haven Navy pistol barrels have a typical New Haven style of barrel with no rib. However, the New Haven Arms Company completed the assembly of the unsold firearms parts from the Volcanic Repeating Arms Company, including a number of Navy pistols which had barrels with ribs.[15]

An 8-inch New Haven Navy pistol, serial number 1673, at top. Below is the rare 6-inch model, serial number 826. The 6-inch model was less popular. Since there was an additional charge for engraving, most New Haven Navy pistols were not engraved.

Top, Edmund Lewis collection, bottom, Bruce Jordan collection

THE NEW HAVEN CARBINES

A 21-inch New Haven carbine with Type II barrel, serial number 100. Virtually all of the silver plating remains on the frame and buttplate. Less than 100 carbines are known in this barrel length. Edmund Lewis collection

Carbines made by the New Haven Arms Company had nominal barrel lengths of 16, 20 and 24 inches and were designated as such in the company's advertising. They are actually 16½, 21 and 25 inches in length. Collectors usually refer to them as 16, 21 and 25 inches. By far, the greatest number produced had 16-inch barrels. Those with 21-inch barrels are very rare, and those with 25-inch barrels are extremely rare. Knowledgeable collectors estimate that less than one hundred 21-inch carbines were produced. Less than twenty of the 25-inch carbines are known today.[16]

The 25-inch New Haven carbine, serial number 73, has a Type III barrel. This is the rarest barrel length with less than twenty known. Edmund Lewis collection

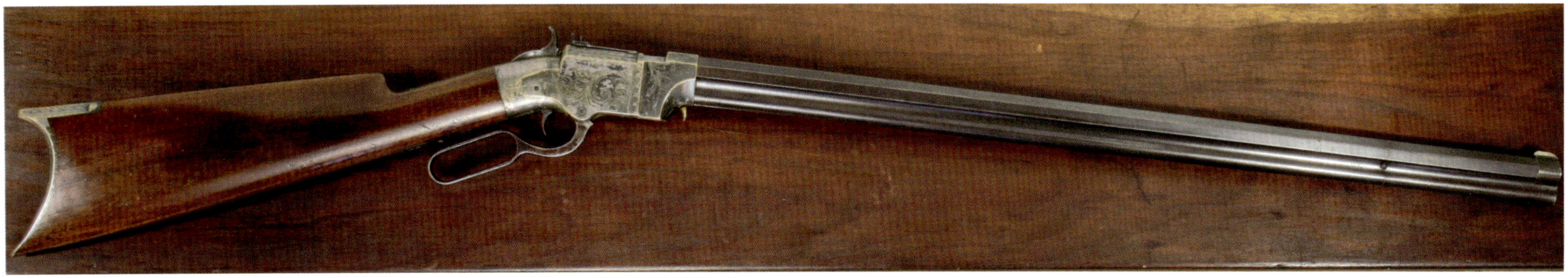

Above: *This 16-inch New Haven carbine, serial number 3062, is the highest serial number known. It has a rare plain polished brass frame. Most were engraved. About 1,000 New Haven carbines in all barrel lengths were produced between 1857 and 1860.* Ray Bentley collection
Below: *21-inch and 25-inch New Haven carbines, both with Type III barrels. Serial numbers 165 and 138.* Donovan Williamson collection

Carbine frames and buttplates were brass and were usually engraved on the top of the frame, the sideplates and upper tang. If the frame was engraved, the buttplate was as well, in a similar design. Frames and buttplates of the carbines were silver-plated at an additional cost of $3.00 as advertised on a circular dated May 1, 1859. One gold-plated carbine is known and appears on page 82. There is a full nickel-plated carbine in the Cody Firearms Museum. Carbines with a plain, polished frame are rare. The levers and carriers were either blued or case-hardened as were the hammers. The levers had a much longer oval to permit the insertion of several fingers. The rear sight was adjustable and the same as found on the Volcanic Repeating Arms Company's pistol-carbine. The standard front sight was a German silver blade inserted into a slot on the magazine collar.

One 16-inch gold-plated carbine is known. Serial number 88. Bruce Jordan collection

A very rare 25-inch New Haven carbine with a Type I barrel, serial number 863. The gun is plated, engraved and has a checkered stock. Collectors sometimes refer to guns with 21-inch or 25-inch barrels as rifles. Ray Bentley collection

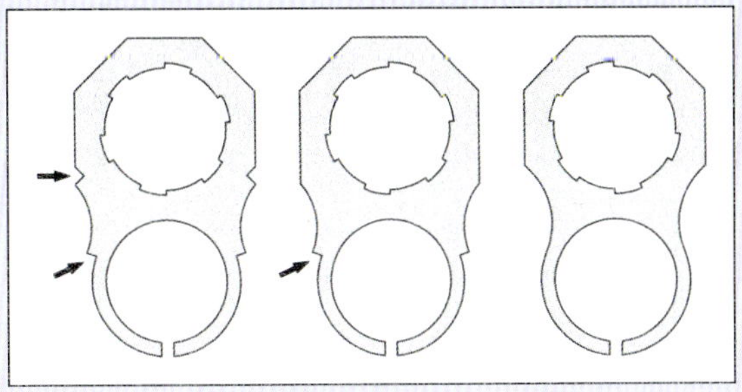

Type I barrel with full rib; Center, *Type II barrel with half rib;* Right, *Type III barrel with no rib.*

Production Numbers — The estimated production in all three barrel lengths is 1,000.[17] The carbines were numbered in the same sequence as the pistols and pistol-carbines. Thus, even though the total production was about 1,000, the highest known serial number on a carbine is 3062.

Barrels and Magazines — There are three distinct barrel types: Type I barrels have a full rib between the barrel and the magazine. This type is thought to have been made up from leftover 16-inch barrels acquired from the Volcanic Repeating Arms Company. They do, however, have a New Haven barrel marking. Two of the highest known serial numbers with Type I barrels, 3039 and 3062, are found on 16-inch carbines.

Type II barrels are transitional and have a half or partial rib that is integral with the barrel but has a distinct junction between the rib and the magazine. Some 21-inch barrels have a partial rib while others do not. These were produced by the New Haven Arms Company, and 16-inch and 21-inch carbines of this type are found.

Type III barrels have a smooth contour with no rib between the barrel and the magazine. This is considered the standard New Haven barrel. 25-inch barrels are known with both Type I and Type III barrels.

The Type II and Type III barrels were made from a single forging with the barrel being rifled and the magazine bored out to the caliber of the carbine. Unlike the Smith & Wesson guns, the rib was decorative rather than for strength. Type II barrels were simplified by omitting one milling operation when only the lower edge was milled. Finally, the smooth contour found on the later Type III barrels required yet less tooling and labor.

More details of barrel dimensions of the New Haven carbines were reported by Thomas E. Hall and are shown in the endnotes.[18] Magazine block and collar dimensions varied depending on the barrel length.[19]

Collectors often refer to the carbines, especially those with 21-inch and 25-inch barrels, as "rifles," but they were called "carbines" in a New Haven advertising circular in May 1859. There is also a reference by the company calling them "rifles" in a later October 1859 company advertisement (shown at left), and many other advertisements also referred to the long arms as "rifles":

NEW HAVEN ARMS COMPANY

The above named company having purchased all the patents on this arm and its ammunition (some eight or ten in number), the inventions of as many of the most ingenious mechanics of the country who have spent years in bringing this wonderful triumph of genius to perfection, are now prepared to manufacture them in a perfect manner, and offer them for sale as the most powerful and effective weapon of defense ever invented. They are made of all sizes, from a four inch Pocket Pistol, carrying six balls, to a twenty four inch Rifle carrying thirty balls.

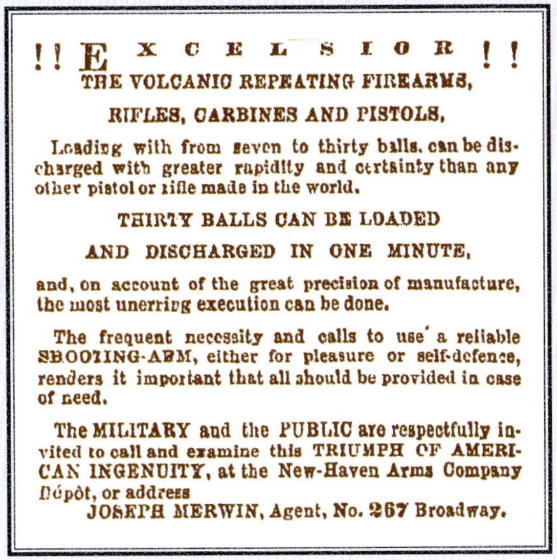

In the 1930s, a number of unsold 16-inch Volcanic carbines, in their original factory boxes, were found stored in the basement of the Winchester Repeating Arms Company. According to Thomas E. Hall, Curator of the Winchester Museum in New Haven, these guns were placed on planks laid over sawhorses and sold at the employee's entrance for fifty cents apiece.[20] How one wishes he could have been there!

THE NEW HAVEN PISTOL-CARBINES

The New Haven pistol-carbine with a New Haven barrel marking is one of the great rarities of Volcanic firearms. When the Volcanic Repeating Arms Company went into receivership, all assets, including various gun parts, became the property of the New Haven Arms Company. Among these were an unknown number of pistol-carbines that were manufactured by the Volcanic Repeating Arms Company but were finished and assembled by the new company. At that time, and in future years, Oliver Winchester was loathe to waste or discard any firearm parts, and when these were assembled by the new company, it was deemed necessary to remove the Volcanic name from the unfinished barrels, leaving only "PATENT NEW HAVEN, CONN. FEB. 14, 1854" in one line. In the example shown at right, only the "O" remains from the first line of the old barrel marking. The original bluing over the obliterated marking indicates that the finishing of the barrel was done afterwards.

These pistol-carbines have a 16-inch barrel; none are known with other barrel lengths. The barrels have a Type I full rib, and serial numbers are found on the lever as on the earlier pistol-carbines manufactured by the Volcanic Repeating Arms Company.

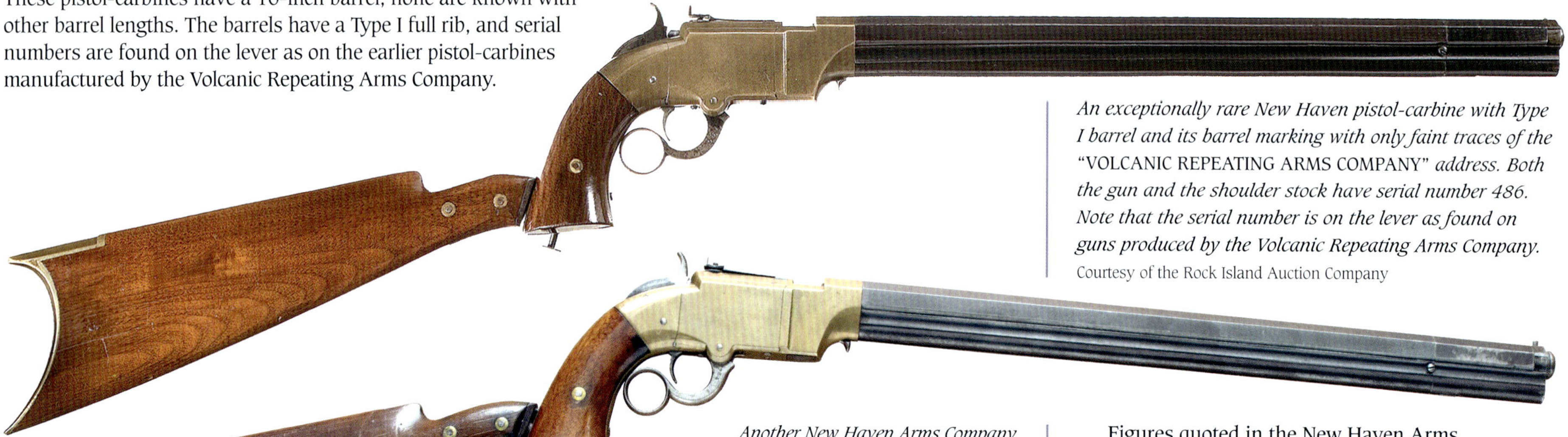

An exceptionally rare New Haven pistol-carbine with Type I barrel and its barrel marking with only faint traces of the "VOLCANIC REPEATING ARMS COMPANY" *address. Both the gun and the shoulder stock have serial number 486. Note that the serial number is on the lever as found on guns produced by the Volcanic Repeating Arms Company.* Courtesy of the Rock Island Auction Company

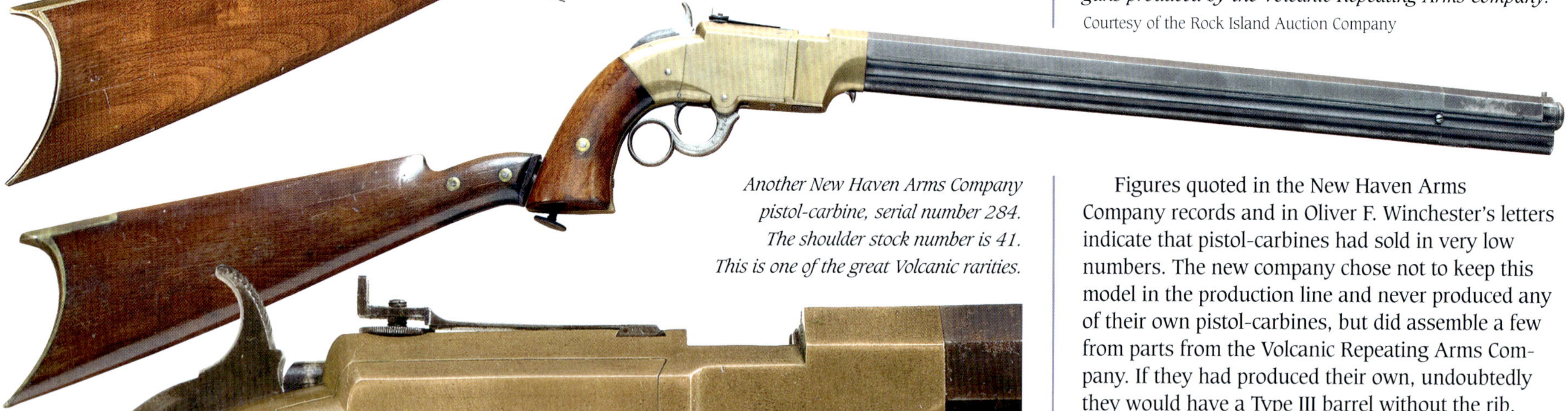

Another New Haven Arms Company pistol-carbine, serial number 284. The shoulder stock number is 41. This is one of the great Volcanic rarities.

The "VOLCANIC REPEATING ARMS CO." *barrel marking on this gun has been obliterated to a greater extent than on the pistol-carbine above.* Stephen Rutter collection

Figures quoted in the New Haven Arms Company records and in Oliver F. Winchester's letters indicate that pistol-carbines had sold in very low numbers. The new company chose not to keep this model in the production line and never produced any of their own pistol-carbines, but did assemble a few from parts from the Volcanic Repeating Arms Company. If they had produced their own, undoubtedly they would have a Type III barrel without the rib.

Collectors should be aware of the existence of New Haven pistol-carbines that are actually made by adding a barrel from a less costly New Haven 16-inch carbine to an earlier Volcanic pistol-carbine. The magazine block, however, of the New Haven pistol-carbine measures 2.265 inches and the collar 0.312 inches, while the dimensions of the 16-inch carbine are 4.500 inches and 0.625 inches respectively.[21] These differences will aid in distinguishing a true New Haven pistol-carbine from one fraudulently assembled with a carbine barrel.

Notwithstanding the expanded inventory, with new pistols and carbines available to the public along with increased promotion and advertising from 1857 to 1860, sales were much less than anticipated. This would soon be changed by the development of the .44 Henry rimfire cartridge and the Henry rifle.

chapter six

Cased and Boxed Firearms

This small-frame Smith & Wesson pistol is the only one known in a cardboard box. Serial number 42D. Perry Hansen collection

Wood-cased and cardboard-boxed pistols and carbines are very attractive and are highly sought after by collectors. They are often in excellent condition since, in many instances, they were not as frequently used. Sometimes these cases also contained one or more boxes of Volcanic cartridges. The 3½-inch small-frame New Haven Pocket pistol is, by far, the most commonly seen pistol to still be found in a wood case.

Pairs of cased pistols are extremely rare as are those of Volcanic pistol-carbines. Cased 8-inch Navy pistols of either the Volcanic Repeating Arms Company or the New Haven Arms Company are seen with greater frequency. Only one wood-cased 25-inch New Haven Volcanic carbine is known. No 20-inch wood-cased or cardboard-boxed carbine has been recorded.

Guns were sold in heavy cardboard boxes covered in black or colored paper or wood-

simulated paper. End or side labels indicated whether the firearm within was plain or plated and engraved. Cardboard boxes are extremely rare due to their fragility. Only one Smith & Wesson pistol is known in its original cardboard box, and only a few New Haven Arms Company Volcanic Pocket pistols are known in original cardboard boxes. Firearms of the Volcanic Repeating Arms Company in their original cardboard boxes are unknown, whether pistols or pistol-carbines. There are a number of 16-inch New Haven Volcanic carbines in wood-simulated cardboard boxes in museum and private collections, but no similar boxes are known containing either 21-inch or 25-inch carbines.

CASED PISTOLS

This elegant cased pair of Smith & Wesson pocket pistols (pictured at right), serial numbers 89E and 53E, is inscribed in a gold panel on the grips "PRESENTED/—BY—/Jos. Patterson". The recipient is unknown. The quality of the engraving, and the full coverage, far exceeds that of other Smith & Wesson pistols and is of a style not found on their other engraved firearms, even on their presentation or exhibition rifles and pistol-carbines. It has been suggested that the pistols were engraved by Richard Bates Inshaw, an English engraver living in Springfield, Massachusetts, at the time.[1] The case, although period, is not typical of other Smith & Wesson wood cases.

Above, left: *Cased Smith & Wesson Type I small-frame pistol with lever spur. Lever spurs were not common but are found on some small-frame pistols in the serial range of 1 to 100 B. Any small-frame pistol with a spur is quite rare.*
Gerald Klaz collection

Left: *Later Smith & Wesson Type I pistol without lever spur, serial number 66D. This was the 466th pistol produced according to the serial numbering system used by the company. This cased pistol is in exceptional condition.*
Gerald Klaz collection

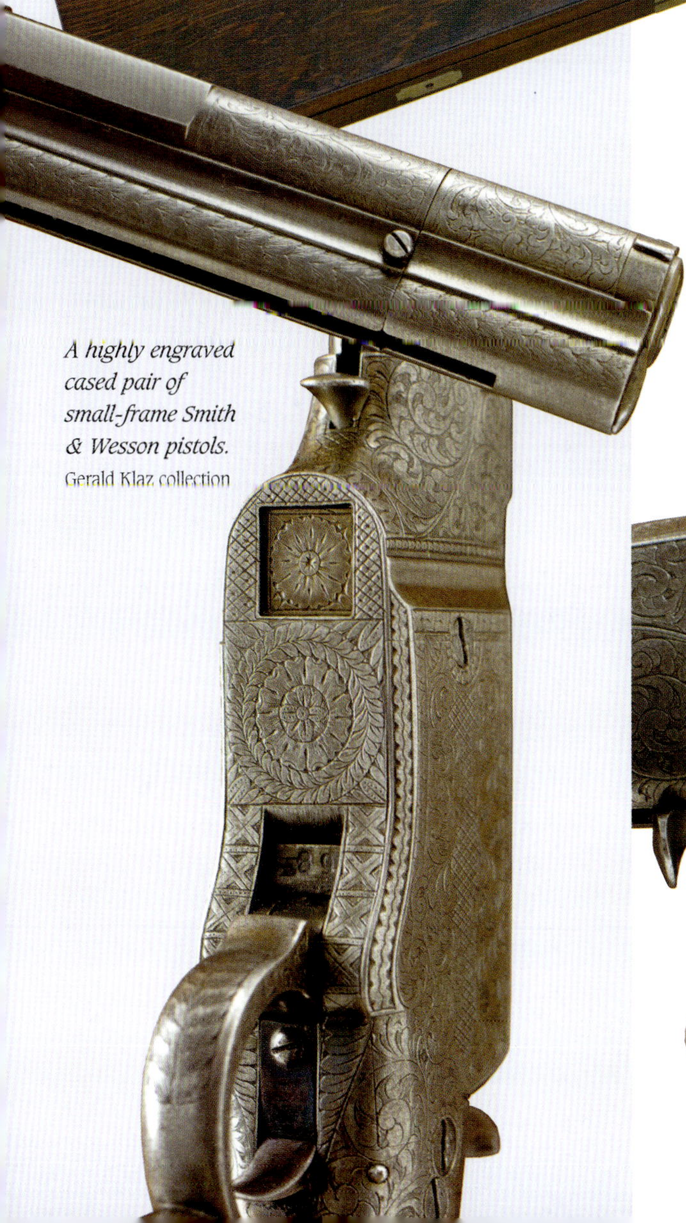

A highly engraved cased pair of small-frame Smith & Wesson pistols.
Gerald Klaz collection

A magnificent Volcanic Repeating Arms Company 8-inch Navy pistol, serial number 1529, cased in walnut.
Courtesy of Rock Island Auction Company

Right: *An early large-frame Smith & Wesson pistol with lever spur, serial number 42, in a period pine casing. The Type I pistol is in superb condition.* Gerald Klaz collection

Below: *This cased Volcanic pistol-carbine has a detachable stock, serial number 414, in .41 caliber, and is one of only a few original cased pistol-carbines. They are extraordinarily rare and are often found with mismatched shoulder stocks. This gun has a shoulder stock that has the same serial number. The case is lined in burgundy velvet and is compartmentalized for the pistol, the stock and a cleaning rod. The cleaning rod compartment now houses an extra magazine spring. The barrel marking is only partially legible. The bolt and carrier have been converted to rimfire.* Courtesy of James D. Julia, Inc.

Above: *A fine Volcanic pistol-carbine cased with a hickory cleaning rod, serial number 452.* Donovan Williamson collection

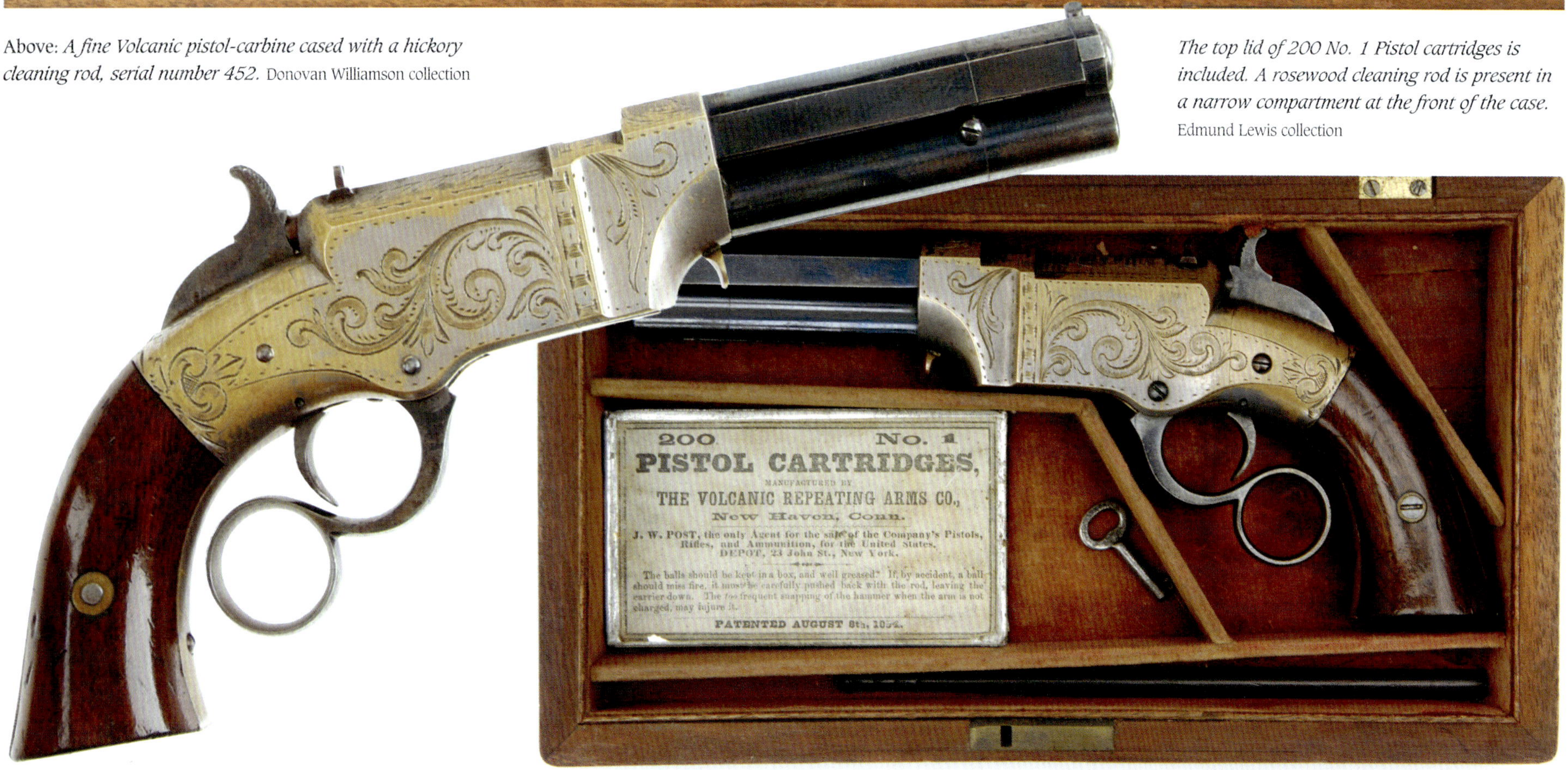

The top lid of 200 No. 1 Pistol cartridges is included. A rosewood cleaning rod is present in a narrow compartment at the front of the case. Edmund Lewis collection

The finest known, unused New Haven Pocket pistol in its original cardboard box. Serial number 1418. Volcanic pistols in this condition are exceedingly rare.
Courtesy of Rock Island Auction Company

New Haven Target pistols are very rare with only 225 having been produced. To find one in exceptional condition and cased, is an extreme rarity. Serial number 1053.
Raymond B. Bentley collection

An early New Haven Pocket pistol in virtually new condition. Serial number 38.
Raymond B. Bentley collection

Right: *New Haven Arms Company 8-inch Navy pistol, serial number 1613, in a mahogany case with green velvet lining. Note the "Y" mark stamped on the lever.*
Gerald Klaz collection

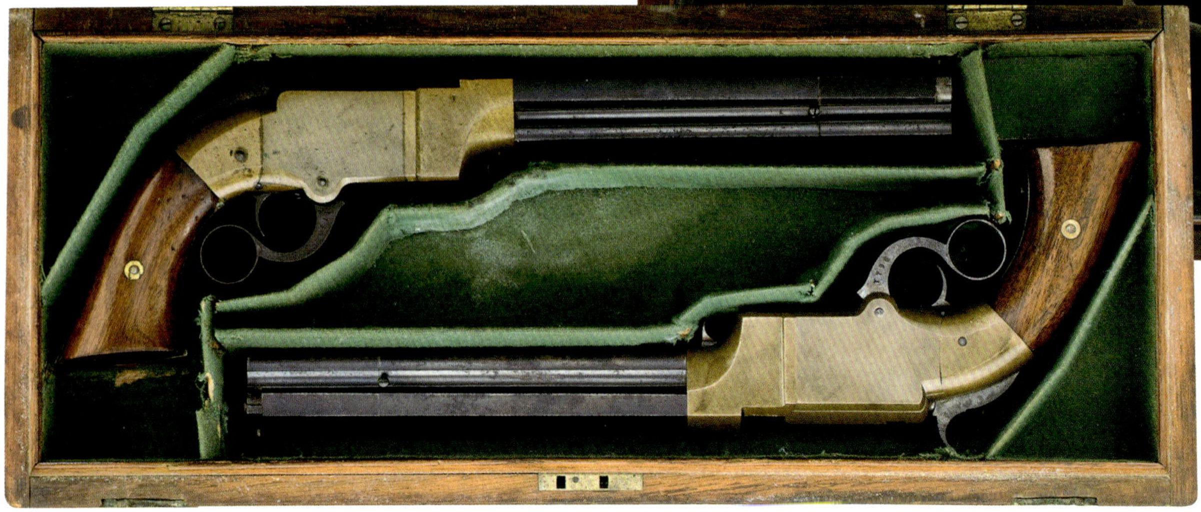

Left: *Exceptionally rare cased pair of Volcanic Navy pistols, serial numbers 7065 and 1796. This English-style wood casing is not compartmented for cartridges.*
Gerald Klaz collection

CASED AND BOXED CARBINES

Wood-cased or cardboard-boxed carbines are extremely rare and very few are known, especially those in a wood casing. Still rare, but more common, are those shipped from the factory in a heavy cardboard box covered with a wood-patterned paper simulating mahogany. Although shipped with two boxes of Volcanic No. 2 cartridges, these are usually now missing. Wood blocks in the upper left and bottom center of the carton helped secure the cartridge boxes. As previously mentioned, Thomas E. Hall, Curator of the Winchester Gun Museum when it was in New Haven, Connecticut, relates that, in the 1930s, the company placed a stack of cardboard-boxed carbines at the employee's entrance and sold them for fifty cents each.[2] In all likelihood, surviving cardboard-boxed carbines were among those.

Carbines in original wood cases are exceedingly rare with only a few known. The authors are aware of only two: one is a 16-inch and the other a 25-inch carbine.

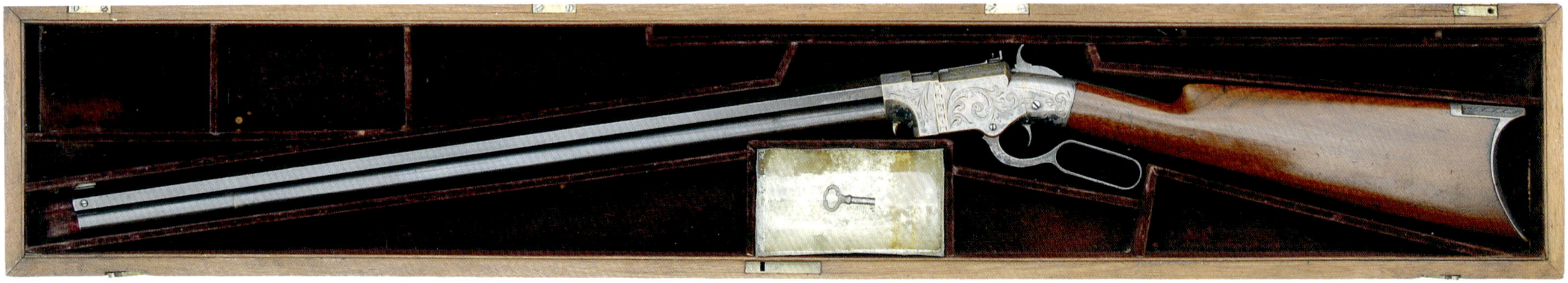

Above and right: *New Haven carbine, serial number 3046. This case originally contained two boxes of cartridges. Note the "PLAIN." label on the side of the box. In Thomas E. Hall's opinion, this was one of the last Volcanic carbines made and perhaps is a good indication of the total number made.*[4]
Courtesy of the Cody Firearms Museum

Left (facing page): *This is the only known wood-cased 25-inch New Haven carbine. Most of its original silver plating remains. Serial number 100.*
Edmund Lewis collection

We know of no 21-inch carbines in either cardboard boxes or wood cases. This may well be a reflection of the very limited production of carbines in these barrel lengths.[3]

These wood-cased and cardboard-boxed firearms are rare and represent the pinnacle of Volcanic firearms available to collectors. Fortunate indeed are those that have these rarities in their collection.

Above: *This new 16-inch carbine was formerly in the Buddy Hackett and Charles May collections.* Courtesy of Rock Island Auction Company

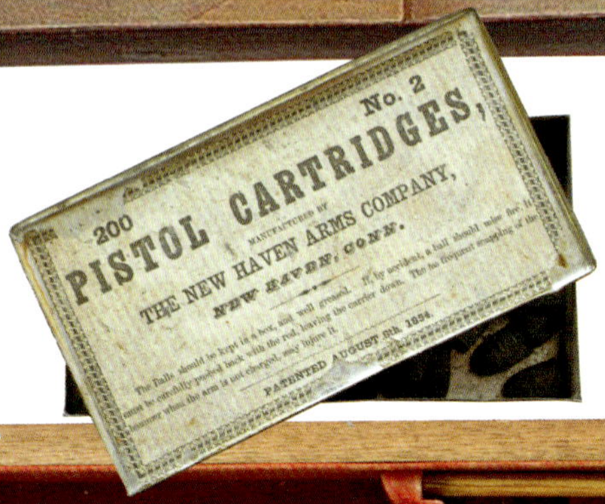

Below: *A wood-cased 16-inch carbine in exceptional condition with a box of No. 2 PISTOL CARTRIDGES. Note the hickory cleaning rod. Cartridge boxes with any remaining cartridges are rare.* Bobby Smith collection

chapter seven

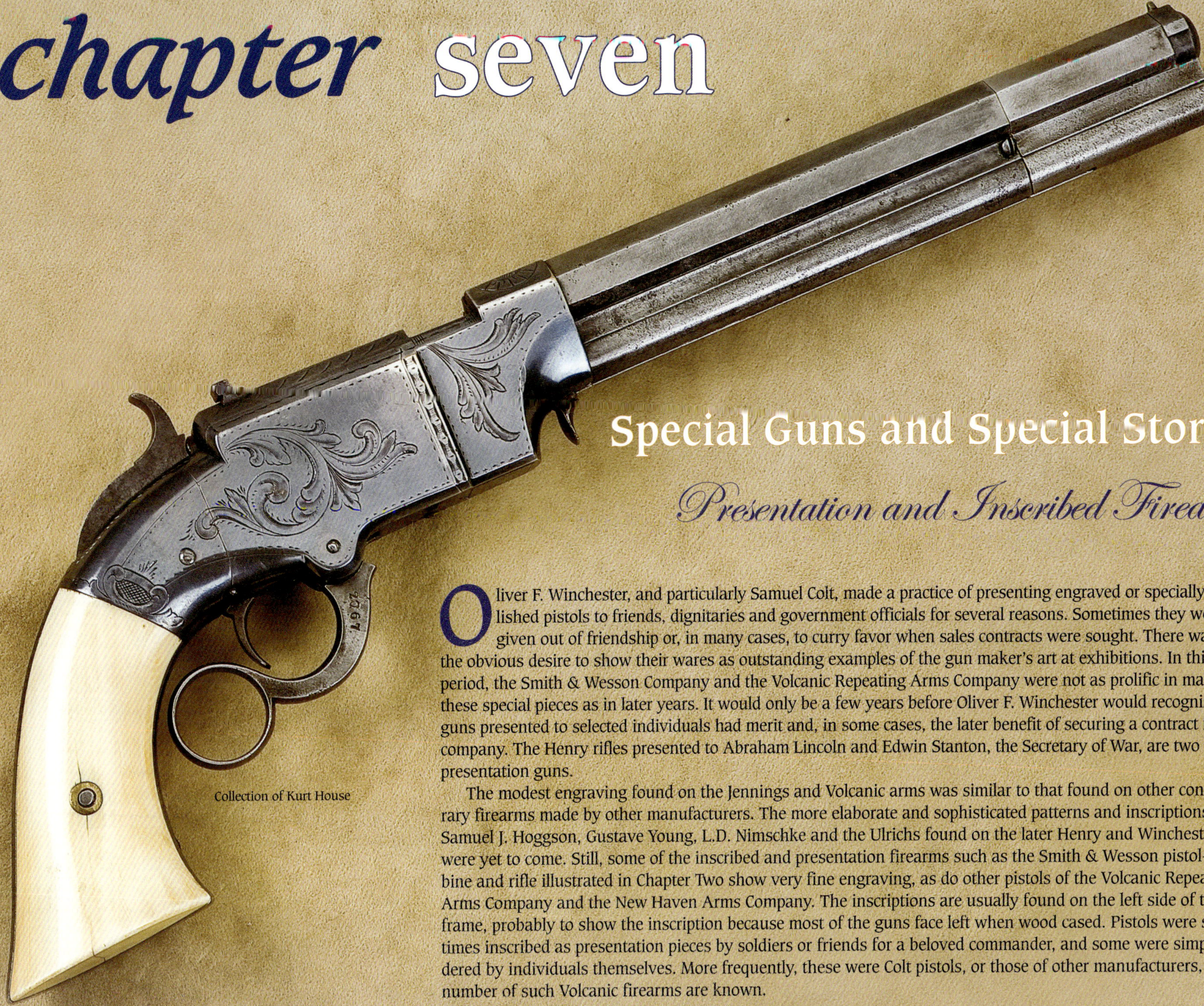

Collection of Kurt House

Special Guns and Special Stories

Presentation and Inscribed Firearms

Oliver F. Winchester, and particularly Samuel Colt, made a practice of presenting engraved or specially embellished pistols to friends, dignitaries and government officials for several reasons. Sometimes they were given out of friendship or, in many cases, to curry favor when sales contracts were sought. There was also the obvious desire to show their wares as outstanding examples of the gun maker's art at exhibitions. In this early period, the Smith & Wesson Company and the Volcanic Repeating Arms Company were not as prolific in making up these special pieces as in later years. It would only be a few years before Oliver F. Winchester would recognize that guns presented to selected individuals had merit and, in some cases, the later benefit of securing a contract for the company. The Henry rifles presented to Abraham Lincoln and Edwin Stanton, the Secretary of War, are two such presentation guns.

The modest engraving found on the Jennings and Volcanic arms was similar to that found on other contemporary firearms made by other manufacturers. The more elaborate and sophisticated patterns and inscriptions by Samuel J. Hoggson, Gustave Young, L.D. Nimschke and the Ulrichs found on the later Henry and Winchester rifles were yet to come. Still, some of the inscribed and presentation firearms such as the Smith & Wesson pistol-carbine and rifle illustrated in Chapter Two show very fine engraving, as do other pistols of the Volcanic Repeating Arms Company and the New Haven Arms Company. The inscriptions are usually found on the left side of the frame, probably to show the inscription because most of the guns face left when wood cased. Pistols were sometimes inscribed as presentation pieces by soldiers or friends for a beloved commander, and some were simply ordered by individuals themselves. More frequently, these were Colt pistols, or those of other manufacturers, but a number of such Volcanic firearms are known.

Collection of Kurt House

AN INVENTOR'S VOLCANIC PISTOL

In 1976, a collector was fortunate enough to acquire a Volcanic pistol that was owned by Benjamin Tyler Henry himself. The pistol, serial number 2067, (shown close-up at left) is a large-frame Navy No. 2 pistol and has the inventor's initials "B·T·H·" inscribed on the left frame. It is engraved and silver plated with ivory grips.

During the period approaching the insolvency of the Volcanic Repeating Arms Company in February of 1857, vigorous efforts were made to promote the company's products in foreign markets including Canada, Europe, Mexico and countries in Central and South America. For this reason, presentation-grade Volcanic pistols are known with the official seals of various countries carved into ivory stocks or engraved on the sideplates.[1]

Pistols, similar to B. Tyler Henry's were presented to Guatemalan President General Carrera in 1856. The present whereabouts of these pistols is unknown.[2]

Collection of Kurt House

OLIVER WINCHESTER'S VOLCANICS

This historic pair of engraved, ivory-gripped pistols, serial numbers 1401 and 1506 (shown at right), was made for Oliver F. Winchester and are the only known firearms to have been owned by Winchester himself.

The engraving is only slightly more elaborate than what is usually found on other pistols. After Winchester's ownership, they remained in the family, belonging to Thomas Gray Bennett, his son-in-law, who had joined the Winchester Repeating Arms Company in 1870 and later became its president.

Unlike Oliver Winchester and William Converse, who both had preceded him as president, Bennett's first-hand experience with firearms and engineering were to serve the company well into the next century.[3] The gold Tiffany & Co. watch also belonged to Bennett. The pistols remain in the family to this day.

Courtesy of R.L. Wilson

A KING'S VOLCANIC PISTOLS

Farouk I became King of Egypt at age sixteen, succeeding his father, King Fuad in 1936. Farouk was extremely wealthy and well known for his glamorous but profligate royal life style. A kleptomaniac by nature, he earned the sobriquet "Thief of Cairo" and, in turn, earned the disgust of his subjects. Along with possessing many palaces and women, he also assembled world-famous automobile, stamp and coin collections. He also enjoyed antique firearms. Included in his collection were two cased Volcanic pistols, which were produced by the New Haven Arms Company. Exiled from Egypt in 1959, Farouk met an untimely end, dying in 1965 at dinner, following an enormous meal.[4]

This extremely rare double casing holds two small-frame New Haven pistols: a 3½-inch Pocket pistol, serial number 1531 and a 6-inch Target pistol, serial number 473. The case is partitioned and lined in purple velvet. The pistols have plain frames, with ivory grips, and are in virtually new condition. Both appear unfired. The case also holds two rosewood cleaning rods, one 6¾ inches long and the other 9½ inches long. The key to the case has a tag written in Arabic, which reads: "from the King Farouk Collection".

King Farouk of Egypt, 1920–1965

Donovan Williamson collection

AN INDIAN FIGHTER'S VOLCANIC PISTOL

George E. Albee was one of those men who had a multifaceted career, excelling in many areas during his lifetime. In June of 1862, at the age of 16, he enlisted as a member of Company G (Wisconsin), 1st U.S. Sharpshooters, organized by Hiram Berdan, and was wounded at the second battle of Bull Run.[5] Following the Civil War, Albee was promoted to 1st Lieutenant in the 24th U.S. Infantry and fought in the West during the Indian Wars. He was awarded the Congressional Medal of Honor for "conspicuous gallantry, meritorious and efficient services… during an engagement with hostile Indians on the 28th and 29th of October 1869 near the head of the Salt Fork of the Brazos River, Texas."[6]

George E. Albee

In 1882, George Albee was employed by the Winchester Repeating Arms Company and played a direct role in the adoption of the Winchester's Hotchkiss Repeating Rifle. He was also a professional marksman for the company, and won the medal at the Creedmoor Matches in Long Island, New York, in 1882 for rapidity and accuracy in shooting.[7] He used a Hotchkiss rifle and .45-70 Govt. cartridges with 600-grain bullets, specially made by Winchester for the match. While at Winchester, Albee took a personal role in modifying firearms for William F. (Buffalo Bill) Cody and Gordon (Pawnee Bill) Lillie for use in their Wild West shows, receiving thankful letters from both men. An article in the *New Haven Register* on January 26, 1891, recounted in great detail Albee's memorabilia and gun collecting activities and specifically mentions his Volcanic Navy pistol. The pistol, serial number 1673, was acquired by Albee in the third quarter of the 19th century along with an extensive collection of other antique arms, edged weapons, military items and documents.[8]

Above: *Brevet Captain George E. Albee wearing his Congressional Medal of Honor. Also shown here are his Volcanic Repeating Arms Company Navy pistol, serial number 1673, a Medal of Honor from the Indian Wars period and an Indian War Medal.*
Edmund Lewis collection

A WOMAN'S VOLCANIC PISTOL

Very few firearms during this period were inscribed to a woman. The only one known to the authors is shown here. It was presented to a Lizzie M. Tomlinson of St. Louis. One fanciful explanation is that she was a madam in a bordello and perhaps it was given as a gift by an admirer for services rendered or for protection in her business pursuits. If walls could only talk, we would perhaps know how the pistol came to be discovered in the New Bedford Hotel in Spokane, Washington, in 1937.

Left: *New Haven Arms Company 6-inch Target pistol, serial number 704. Engraved "Lizzie M. Tomlinson./St. Louis."*
Courtesy of The Northwest Museum of Arts & Culture, Spokane, Washington

Right: *A bordello clock, circa 1885. Unfortunately, the woman's hat and the Volcanic pistol cover a riband at the base of the statue that reads: "TIME IS MONEY".*
Bust courtesy of David Zarneskie

NEW HAVEN ARMS COMPANY CARBINE SERIAL NUMBER "0"

In February of 2008, the estate auction of Edward Martin, the governor of Pennsylvania from 1943 to 1947, advertised that, among his small firearms collection to be included in the auction was a Volcanic carbine, serial number 36. Martin had died in 1967, and his son had kept the estate intact until his death in 2008. The remaining heir, a distant niece, knew nothing of the history of this carbine. Upon contacting the auction house, it was learned that the "36" was stamped on the undersurface of the receiver and was most likely a rack number. When the auctioneer was asked to check for the true serial number, it was found in the expected internal locations of the brass buttplate and wood of the stock beneath it, the left side of the tang and just beneath the head, on the shank, of each screw. What was unexpected was that the serial number was a "0".[9]

Martin was born in 1879 and had a very distinguished military career. He served during WWII as a Major General in charge of training the troops from Pennsylvania prior to his election as governor in 1943.[10] In addition to this 21-inch New Haven Arms Volcanic carbine, his small collection had consisted of a Henry rifle and three more modern military rifles. The barrel marking is "NEW HAVEN CONN. PATENT FEB. 14, 1854". Madis states that variations in New Haven Arms Company barrel markings occur, though infrequently, particularly on pistols and carbines with lower serial numbers, where the marking maybe just one line, the company name preceding the patent line, as is the case with this carbine.[11]

The mystery of a "0" serial number will likely remain enigmatic, and the most frequently suggested explanation is that it was a prototype carbine.

Above: *Volcanic carbine, 21-inch Type II barrel, serial number "0".* Stephen Rutter collection

Buttstock, screws and lower tang all marked with serial number "0". Stephen Rutter collection

BENJAMIN TYLER HENRY'S NEW HAVEN PISTOL-CARBINE

This unique New Haven pistol-carbine, serial number 1467, is .22 caliber and was found at a barn sale in Claremont, New Hampshire, by the well-known collector Eldon Owens. It is thought to have been made by B. Tyler Henry for his grandson. Henry was originally from Claremont, and Henry's great-grandson owned the farm on which the sale was held.

The frame is smaller than on other pistol-carbines and is engraved in a pattern unlike that found on other Volcanic arms. The barrel is part-octagon, part-round with the octagonal portion having a Damascus finish. The round part is rust brown and has an octagonal ring at the rear which retains and acts as a locking device for the magazine cover. The bolt head is just wide enough for a .22 caliber BB cap. This unique firearm has a steel skeleton stock that attaches to the grip with a pin and screw as on other Volcanic pistol-carbines. When found, the pistol was in the holster shown. The extremely rare brown leather holster is 26 inches long and has a closing strap.

Despite these and other attempts to promote sales of the Volcanic firearms, their deficiencies, and the deficiencies of the ammunition they used, continued to prevent the companies from achieving sales sufficient to make these enterprises profitable. They were, however, innovative and attractive firearms to people living in the 1800s as well as to many of today's collectors.

This New Haven pistol-carbine has a distinctly different engraving pattern than seen on other New Haven arms. It was in the holster shown here when found.
Courtesy of James Julia, Inc.

Engraved New Haven Target pistol, serial number 553. Edmund Lewis collection

chapter eight

The Volcanic in Service

A resolute soldier of the Civil War holding a small-frame New Haven Pocket pistol and a Bowie knife. Casey Nanz collection

Few stories of the actual use of Volcanic firearms are known, and it is recognized that they were not often carried as side arms by soldiers in the Civil War. In a few instances, Volcanic carbines were pressed into service by the United States Army as well as civilian law enforcement agencies. Others were carried by western adventurers and explorers.

CIVIL WAR VOLCANICS

Unquestionably, Colt pistols and those of other manufacturers, were used as side arms by Civil War soldiers. These pistols were usually of government issue, but were often purchased by themselves or given to them by friends. Instances of the use of Volcanic arms are sometimes documented, either by contemporary written accounts or by daguerreotypes or tintypes. Occasionally, guns — invariably pistols — are found with inscriptions indicating usage during this conflict.

This New Haven Pocket pistol was carried during the Civil War by Thomas Howard. National Archives military and pension records indicate that he served with the 61st New York Volunteers as well as the 44th New York State Volunteers and was present at the battles of Chancellorsville and Gettysburg. He was injured near Snickersville Gap, Virginia, but continued in service until the end of the war.[1]

The backstrap of this small Pocket pistol is engraved "T. HOWARD. 61st Co. D." Courtesy of Little John's Auctions

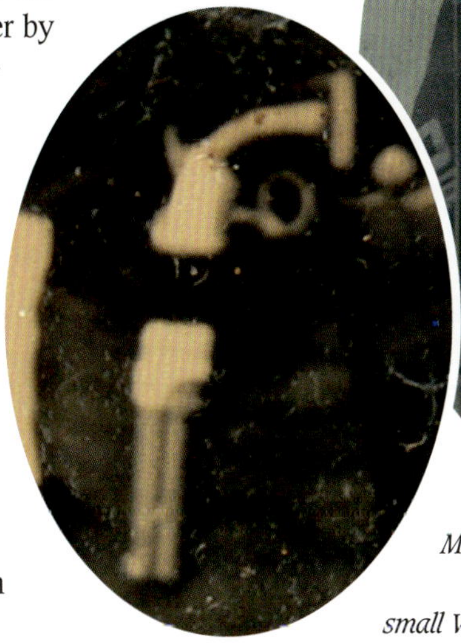

Melaniotype of Civil War 1st Infantry Sergeant. Note close-up at right of small Volcanic Pocket. Frank Graves collection

This Volcanic pistol is inscribed on the frame: "WH CADWELL/Brewsters Rifle Co/3'd Regiment C.V."

Union soldier with a 6-inch Navy pistol in his belt and Allen & Wheelock pistol in hand. Donovan Williamson collection

Tintype showing a Union soldier with a musket and a Pocket pistol in his belt. Donovan Williamson collection

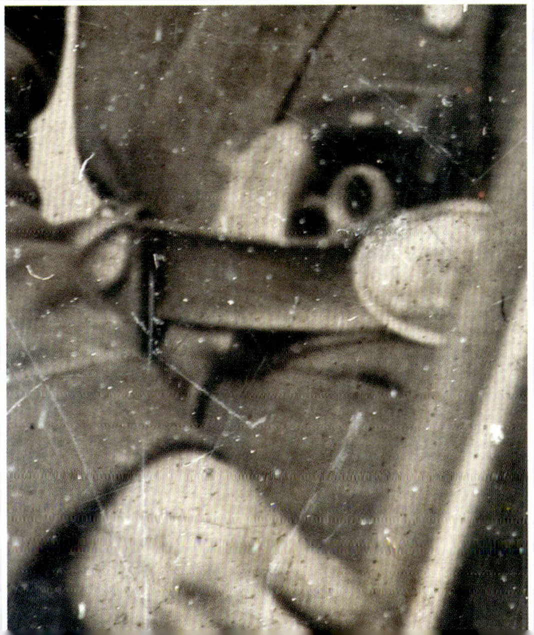

The Third Connecticut Volunteers were among the first to answer Lincoln's call in April of 1861. As part of the Army of the Potomac, it led the column on the march towards Richmond and later served with great courage at the first Battle of Bull Run.

A LIEUTENANT'S VOLCANIC

The left frame is inscribed in period script and block letter engraving: "Lieut. Thos Gardner/From/ SOME OF THE "RANGERS"/&/"ORIGINAL CROTONS."/March 12th 1857. The right sideplate is inscribed: "Keep this weapon "TOM."/Never USE it in a BAD CAUSE./Never SURRENDER it in a GOOD./Remember that FREEDOM is worth more than LIFE/That FREEMEN never can be SLAVES."

Gardner was apparently a member of the "Rangers" who surely were "free-staters" participating in the violent struggle taking place in Kansas and Missouri in the period preceding the Civil War.

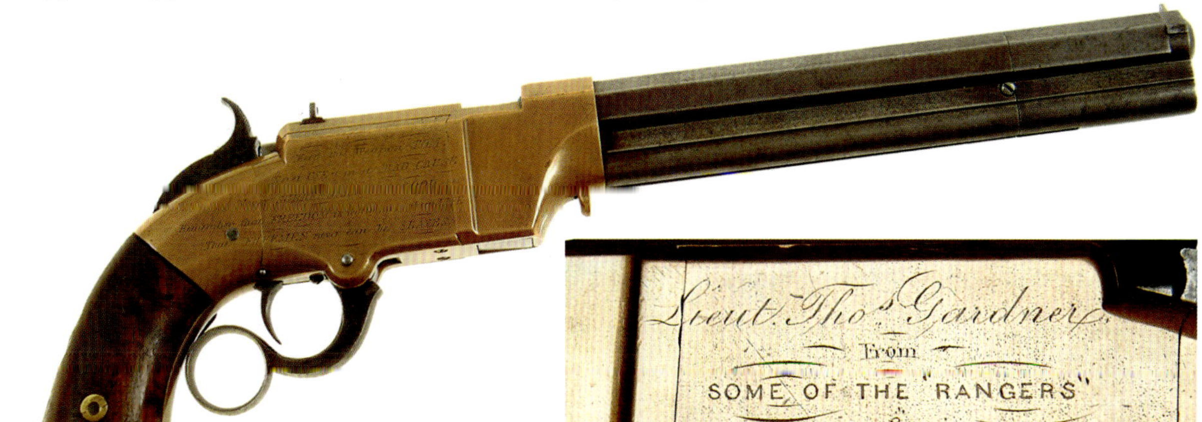

This 8-inch Navy pistol, serial number 1080, was presented to an Army Lieutenant in 1857. Courtesy of James D. Julia, Inc.

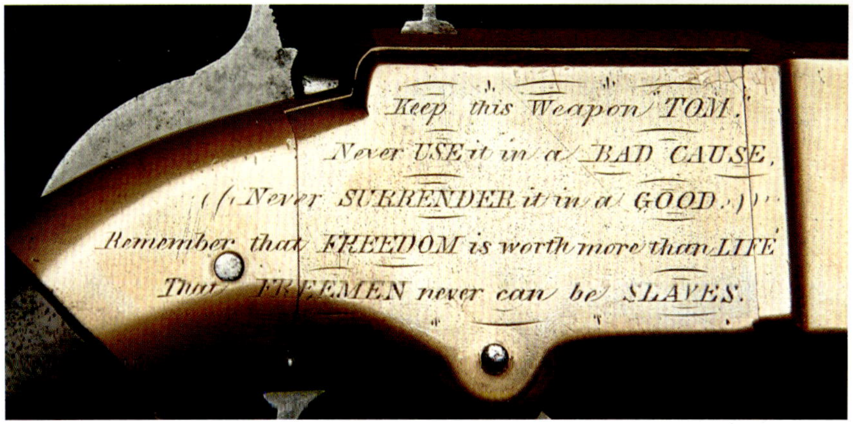

It is uncertain, but the "Crotons" may have had a connection to a university fraternity.

Thomas W. Gardner enlisted as a Captain in Company G, Ohio, 90th Infantry until he resigned on May 5, 1863. He died in Missouri in 1918.[a]

MAJOR ROCKWOOD'S PISTOL

Boxed pistols of this quality are extremely rare.
Alden S. Bradstock, III collection

This very fine New Haven Pocket pistol has virtually all of the silver plating intact and is contained in an excellent cardboard box with an end label stating the pistol was "PLATED & ENGRAVED".

Theodore H. Rockwood,
from the
First Baptist Choir,
New Haven.

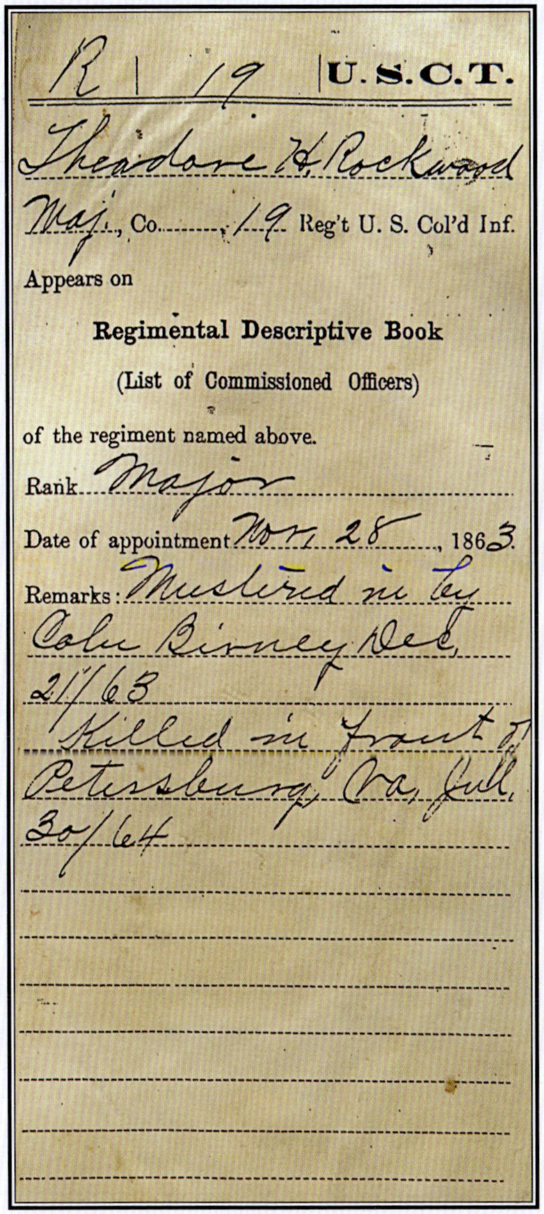

R | 19 | U.S.C.T.

Theodore H. Rockwood

Maj., Co. ____, 19 Reg't U. S. Col'd Inf.

Appears on

Regimental Descriptive Book

(List of Commissioned Officers)

of the regiment named above.

Rank Major

Date of appointment Nov. 28, 1863.

Remarks: Mustered in by Col. Birney Dec. 21/63

Killed in front of Petersburg, Va., Jul. 30/64

Regimental Descriptive Book (List of Commissioned Officers) of the 19th U.S.C.T., listing Major Rockwood as "Killed in front of Petersburg, Va., Jul. 30/64".

Alden S. Bradstock, III collection

Theodore H. Rockwood was born in New Haven, Connecticut, in 1836 and as a young adult received, from the Choir of the First Baptist Church, a presentation pistol made by the New Haven Arms Company. Just why he received this gift is unknown. Perhaps it was because he was mustered into the service of the United States in Company E of the 1st Connecticut Heavy Artillery on May 21, 1861, during the Civil War. He was given the rank of First Lieutenant and served in that unit before he transferred to become a Major of the Infantry in December of 1863. He joined the 19th Regiment of U.S. Colored Troops (U.S.C.T.) as a staff officer and trained with the troops at Camp Stanton in Benedict, Maryland. The unit was attached to the Fourth Division of the Army of the Potomac from April until July of 1864, when it found itself in front of the Confederate defenders around Petersburg, Virginia.

The rebels were deeply entrenched and, in an attempt to break the stalemate, the Union forces devised a plan to build the longest underground mine in military history. The tunnel was 511 feet long, allowing the Union forces to detonate 8,000 pounds of gunpowder under the Confederate breastworks. The explosion killed many defenders and created a huge crater. The Confederate defenders recovered, and the "Battle of the Crater" followed with the Union soldiers attacking across the canister swept "No Man's Land." Colonel Henry G. Thomas tried to rally his men after getting pinned down on the western edge of the crater. Thomas saw officer after officer fall dead and said of one officer, Major Theodore H. Rockwood of the 19th U.S.C.T., that he "mounted the crest of a trench and fell back dead with a smile on his lips." Had Major Rockwood stayed with the Heavy Artillery and not transferred to the Infantry to get a promotion, he may have survived the Civil War.[3]

MAJOR GENERAL BUTLER'S GREAT SOUTHERN EXPEDITION

The vanguard of Major General Benjamin Franklin Butler's southern expedition to capture and occupy the Confederate port of New Orleans, Louisiana, sailed from Fortress Monroe, Virginia, on the war steamship U.S.S. *Constitution* in September of 1861. The ship carried the 9th Connecticut and 30th Massachusetts Regiments along with Captain Reed's squadron of Mounted Rangers, carrying Volcanic rifles along with side arms and sabers.

General Benjamin Butler was one of the most controversial generals of the Civil War. His military career was checkered, with some important successes and some failures. He commanded the Great Southern Expedition, himself embarking on the steamship *Mississippi* from Fortress Monroe in Hampton, Virginia.

Butler sailed southward capturing Forts Hatteras and Clark in North Carolina in 1861 and later directed the first Union expedition to Ship Island, off the Mississippi Gulf Coast. He went on to occupy New Orleans in May of 1862 after its capture by Admiral David Farragut.

Benjamin Franklin Butler

DEPARTURE OF THE GREAT SOUTHERN EXPEDITION, UNDER GENERAL BUTLER, FROM FORTRESS MONROE. Harper's Weekly, September 14, 1861

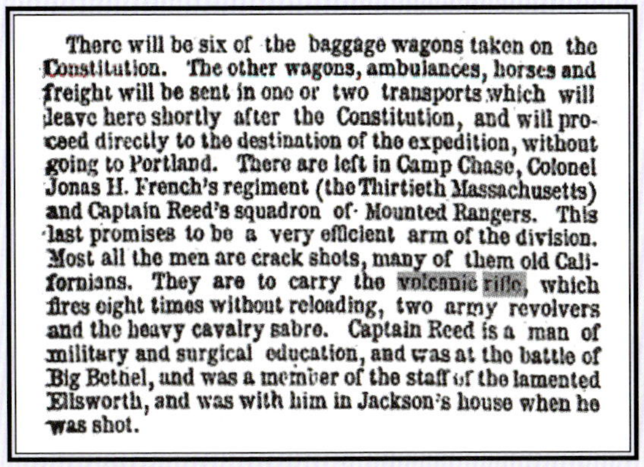

There will be six of the baggage wagons taken on the Constitution. The other wagons, ambulances, horses and freight will be sent in one or two transports which will leave here shortly after the Constitution, and will proceed directly to the destination of the expedition, without going to Portland. There are left in Camp Chase, Colonel Jonas H. French's regiment (the Thirtieth Massachusetts) and Captain Reed's squadron of Mounted Rangers. This last promises to be a very efficient arm of the division. Most all the men are crack shots, many of them old Californians. They are to carry the volcanic rifle, which fires eight times without reloading, two army revolvers and the heavy cavalry sabre. Captain Reed is a man of military and surgical education, and was at the battle of Big Bethel, and was a member of the staff of the lamented Ellsworth, and was with him in Jackson's house when he was shot.

The New York Herald, November 21, 1861.

Pictured at left was a newspaper article from *The New York Herald* dated November 21, 1861: "There will be six of the wagons… Most of the men are crack shots, many of them old Californians. They are to carry the Volcanic rifle, which fires eight times without loading, two army revolvers and the heavy cavalry sabre." Butler's administration of the city was stern but fair, though some of his acts regarding the treatment of women, who despised the Union soldiers, gave great offense both in the North and South, as well as in England and France.[4] He earned the sobriquet "Beast Butler" from the southern women and "Spoons" Butler for his alleged habit of pilfering the silverware of homes in which he stayed.

Recalled to Washington, Butler was given command of the Army of the James. Always innovative, he saw great promise in the use of Richard J. Gatling's 1862 invention, the Gatling gun, and was the only Union general to make use of it. Butler managed to obtain seventeen of these weapons, which could fire 250 rounds a minute, but the Union high command ignored the weapon completely.[5] Military failures late in the war and political considerations by Lincoln, Ulysses S. Grant, Edwin Stanton and others resulted in the end of Butler's career as a general.[6]

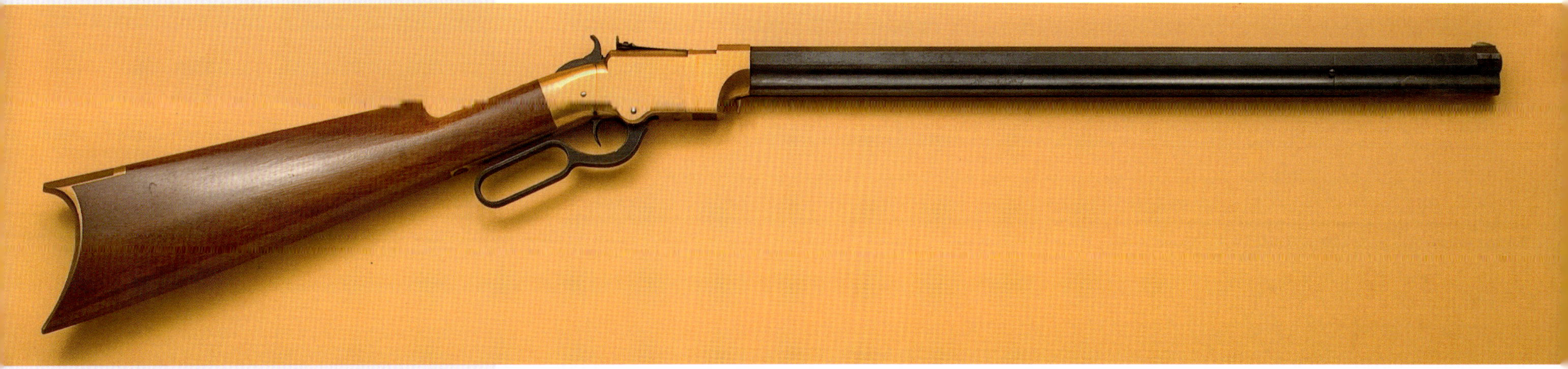

In all likelihood, 16-inch carbines were carried by the Mounted Rangers.

THE STATEN ISLAND "YELLOW FEVER" WAR OF 1858

In his authoritative article, "The Volcanic in Service — During the Staten Island War of 1858," which appeared in the September/October 2005 issue of *Man at Arms* magazine, Herbert G. Houze relates how Volcanic carbines were used by the New York City police in quelling riots arising from public anger directed at the Yellow Fever Hospitals on Staten Island in New York harbor.[7]

Yellow fever is an acute viral hemorrhagic disease transmitted by infected mosquitoes found in the tropical and subtropical areas of South America and Africa. In the mid-19th century, prior to understanding the true vector of transmission, contagion was commonly believed to occur by breathing the "horrible vapors" of infected patients. Seaport cities often established quarantine hospitals to house the sick as well as those who had been exposed to the disease. The New York City Quarantine Station was located on 30 acres of land on Staten Island that had been sold by Daniel D. Tompkins, a former Governor of New York and Vice President of the United States, with the provision that if its use for that purpose was abandoned, the property would revert to the seller or his heirs. Over the years, the once distant towns grew closer to the Quarantine Station, and concerns and hostility reached the boiling point.

On the night of September 1, 1858, a mob, organized by Ray Tompkins, dragged patients from their beds and put them outside in the rain prior to burning the Men's Yellow Fever Hospital, as well as the Small Pox Hospital and other buildings. Ray Tompkins, chief of the Fire House in nearby Tompkinsville, was not acting entirely for the public good, inasmuch as it was he that stood to benefit when the property reverted to the donor's heirs.

On September 2nd, a small contingent of U.S. Marines was posted to the Federal Government buildings and, believing that the worst was over, city authorities delayed the transfer of additional police to Staten Island. Seeing the opportunity to finish their work at about 11:00 p.m. that night, Tompkins and his gang brutally attacked the remaining buildings, burning the Women's Yellow Fever Hospital as well as the homes of Doctors Bissell and Walles. The City and State of New York, however, did not abandon the Quarantine Station and immediately erected tents to house not only the displaced patients and staff, but also for a protective armed garrison. As published in *The New York Times* on September 6, 1858, "The police force at Quarantine consists of 100 men... Each man is armed with a Volcanic repeating rifle, loaded with twenty balls, which can all be fired in less than two minutes, the aggregate is equal to one thousand shots." Gradually, through the fall of 1858, the tensions eased. Ray Tompkins was acquitted at trial, claiming that he had acted not out of malice but only on the public's behalf.

Beginning on October 12, 1858, Joseph Merwin, the New York City agent for the New Haven Arms Company, capitalized on the police's use of the Volcanic firearms, promoting them by placing banner advertisements in *The New York Times*. In April of 1859, the concerns of the citizens over the rebuilt Quarantine Station were renewed, and once again the posting of police with "Volcanic rifle" firepower was reiterated in the news, and the crisis was averted. The Quarantine issue was defused in June of 1859 when a steamer ship, the *Falcon*, was converted to a hospital ship and moored in the approaches to New York harbor.

According to newspaper reports, 120 Volcanic carbines saw service in the hands of the New York City Police during the "Staten Island War." Because of the reference to their being "loaded with twenty balls," it is reasonable to assume that they were 16-inch carbines. On May 26, 1859, *The New York Times* reported, "The bill of the New Haven Arms Company, amounting to $1,150, for Volcanic arms which the Company let the authorities have during the Quarantine riots, in September last, was received." Based upon the numbers in the article, Houze estimates that approximately 50 were kept in service and 70 were returned to Merwin. The fate of these Volcanic carbines is unknown, but based upon a November 7, 1890, document in the Winchester Archives, Houze postulates that, "41 old carbines for 5 RSG 10ga. 20in" in the document refers to these Staten Island Riot Volcanic carbines.

Above, left: *Identification number "19" stamped on the receiver of a New Haven Arms Company 16-inch carbine, serial number 136, pictured below. This carbine, and others noted with similar stamped numbers, are thought to be among those stamped with inventory numbers by the New York Metropolitan Police.* Stephen Rutter collection

Above, right: *Another "Yellow Fever" carbine, number 8.* Courtesy of Rock Island Auction Co.

New Haven Arms Company 16-inch carbine, serial number 136. Stephen Rutter collection

Houze further states that the document could explain the origin of the 35 to 50 Volcanic carbines given by the wife of the president of the Winchester Repeating Arms Company to the Calvary Baptist Church Boys' Brigade in 1893. They were intended to be used by the drill and marching team, and were deemed suitable for the boys because of the carbine's diminutive size. Originally, the stamped numbers on the bottom of the receivers, forward of the finger lever pivot, of these Volcanic carbines was presumed to have been placed as "Regimental Numbers" by the church. But the sequence of numbers identified with these Volcanics ranges from 14 to 96, well beyond the number given to the church. The sequence range would, however, make sense if they had been numbered before they were given to the Calvary Baptist Church. This conundrum, coupled with the "41 old carbines" document and the likelihood that the New York City Metropolitan Police Department used some external mark to keep track of the 120 Volcanic carbines originally in their possession in 1858, permits speculation that these guns are one and the same. It is now believed, however, that rifles were marked in this manner while they were in use by of the New York City Metropolitan Police.

A Proclamation!
THIRTY MILLION DOLLARS SAVED TO THE COUNTRY.
THE PRESIDENT NO LONGER THINKS OF BUYING CUBA.

Argument and entreaty are exhausted. He has now come to a very sensible conclusion, viz.: to arm one thousand men with the VOLCANIC REPEATING RIFLE, which carries thirty balls, and can be loaded and discharged thirty times in less than one minute. One thousand men, armed with this Rifle, would be equal to thirty thousand armed with any other weapon. If our worthy *Chief Magistrate* does not bestir himself the Cubans will be ahead of him, as they are now buying them in large quantities, and will soon be able to defend themselves against any force that can be brought against them. The balls are all waterproof—hence their great superiority over any other arm. No changes of climate, or even immersion in water, affects them.

Colt's Pistol, so long considered the *neplus ultra* of Firearms, is now fairly outdone by this new invention. Their beauty of construction and simplicity of movement are the wonder and admiration of all who have seen them. As was remarked a few days since, by one of our most accomplished commanders, "The man who invented that gun was a benefactor to his race, and the nation that was in possession of such a weapon need never fear even the combined armies of the world, if their cause was just."

Guns and Pistols, carrying from seven to thirty balls. This arm is now on exhibition and for sale, at wholesale and retail, at the Company's dépôt, No. 267 Broadway, New-York.
JOSEPH MERWIN,
Sole Agent for United States.

Also agent for the celebrated G. D., C. & T., and Waterproof Percussion Caps.

"He [Buchanan] *has now come to a very sensible conclusion, viz.: to arm one thousand men with the VOLCANIC REPEATING RIFLE, which carries thirty balls, and can be loaded and discharged thirty times in less than one minute. One thousand men, armed with this Rifle, would be equal to thirty thousand armed with any other weapon. If our worthy Chief Magistrate does not bestir himself the Cubans will be ahead of him, as they are now buying them in large quantities..."*
The New York Times, February 5, 1859

ON TO CUBA!

Volcanic firearms were also pressed into service when there appeared to be the prospect of invading or acquiring Cuba in 1859. The Hew Haven Arms Company's aggressive agent Joseph Merwin had this advertisement (shown at left) placed in *The New York Times* on February 5, 1859, in which he suggested that President James Buchanan should invade and annex Cuba.

The fervor to abolish slavery after John Brown's raid on the armory at Harpers Ferry in October of 1859 was rampant, and there was considerable sentiment for the invasion of Cuba to end slavery there as well. While the United States Senate was considering whether to try annexation of the country, Richard H. Dana, author of the American classic *Two Years Before the Mast,* traveled there extensively and found favor with this concept.[8, 9] President Buchanan obviously thought little of this idea, and it never happened.

A "SMALL, BUT DEVILISH" VOLCANIC

Herbert G. Houze has kindly provided this interesting account of a man's experience with a Volcanic pistol in the Levant:

"Without doubt one of the staunchest proponents of the Smith & Wesson Volcanic pistol was William Cowper Prime (1825–1905). In turn, a successful attorney, traveler, publisher and art historian, Prime spent most of the years 1855 and 1856 in the Middle East. Upon returning to New York, he published two accounts of those travels. In the first work, *Boat Life in Egypt and Nubia* (Harper & Brothers; New York: 1857), Prime gave the following advice to those contemplating a similar trip:

"Arms are useful only for show in Egypt. It is well to have a good pistol, and in Syria it is necessary. I found the volcanic pistol, as it is called, much preferable to Colt's. The latter was constantly getting out of order, and from the falling of the cap between the cylinder and the hammer, was useless twice out of three times after the first shot. I carried the fixed ammunition of the volcanic pistol with me in all climates, and found it infallible. It is compact and safe. I recommend it to the eastern traveler."[10]

In the second volume chronicling Prime's tour of the Levant, *Tent Life in the Holy Land* (Harper & Brothers; New York: 1857), he repeated and expanded upon this advice:

"The traveler who purposes [sic] *visiting Syria will need to provide himself with good pistols. I recommend the volcanic pistol, formerly known as the Jennings' patent, as altogether preferable to any that I have seen. It is light, safe, and sure, and the ammunition compact and easily carried."* [p. 496]

It should be pointed out that Prime's endorsements were not idly made. In *Boat Life in Egypt and Nubia*, he recounted a confrontation with Arabs at the Tomb of Apis near the pyramids of Giza. When the leader of this group threatened to do him bodily harm, Prime seized the man and pressed "an ugly-looking Volcanic pistol at the side of his head." However, it was during his travels to the Holy Land and Syria that Prime found the most use for his Volcanic pistol. He recorded that, when he left Jerusalem, he was armed with "a navy Colt in one fold of my shawl and a volcanic repeater in the other (small, but devilish), not to mention a bowie-knife."[11] These precautions proved wise when Prime and his traveling party entered Hebron. Confronted, while on horseback, by a stone-throwing mob, Prime and his companion drew their revolvers:

"The shudder of fear that went through the assembly was actually visible. Two balls in that mass would have made terrible havoc. We stood thus, facing each other for thirty seconds, and then I raised my pistol and fired six balls successively in the air, and replacing it in my shawl took out another (my volcanic repeater) and shook it in their faces to be a lifesaver.

It was enough. The terror of that revolver will remain in Hebron as a civilizer. We turned our horses' heads again and rode very slowly on, but no more stones were thrown, nor did one of the crowd pass that spot."[12]

This event and the constant watch for bandits would later cause Prime to draw his Volcanic pistol again. While traveling through Syria, Prime was awakened by an approaching storm about two o'clock in the morning. To his utter dismay, upon leaving his tent, he found that the camp's guards were asleep. After rousing them, he surveyed the area where they were staying and made a startling observation:

"There was one rocky point that I had noticed before dark as commanding our position most beautifully, and on this I fixed my eyes now more intently than elsewhere. Was it imagination, or did I see a moving object on the surface of that flat rock? The moonlight certainly shone on something, though I could not see distinctly what is was, that it was not rock. If it were a man why did he now not drop me? He had a beautiful shot as I stood out in the black boor noose against the white tent. I was not altogether easy at that instant. I had the sensation of an entering bullet in my throat, breast, brain — I couldn't exactly tell where it would hit — But I felt it somewhere generally, as a Hoosier might say.

At all events I could not go to sleep with that fellow lying on the rock, and now I began to reason on this wise. If he is a man he must be an enemy. A friend would have no business there, and a villager would not be there. He can't be there with reference to anyone else, for his position commands our tents and not the village. He must be watching us, and if so, it is for no good — and as I reasoned I had gone into the tent, taken out my small volcanic pistol which carries a ball an immense distance and was much preferable to Colt's for sharp shooting and returned to the front of the tent. I continued my reasoning. Shall I send a ball up at him? If I do I must hit him, or he will hit me, as certainly as I fire. In short, I must bring him down, so here goes, and I threw up my pistol and I sent a conical ball, whistling as those hollow balls always go..."[13]

Prime's target, however, was not a man. Rather it was a gray wolf waiting its chance to sneak into camp to snatch some discarded soup bones. Nevertheless, Prime's shot not only mortally wounded the beast (it was formerly dispatched with a knife by the cook), but also proved the Volcanic's effectiveness. This latter point is worth remembering, since many today think of the Volcanic and its "Rocket Ball" ammunition as little more than novelties. Mr. Prime certainly was a robust fan of the Volcanic, and one wonders, suspiciously, if he was a friend of Oliver Winchester, but perhaps that is unfair. What Samuel Colt would have said about Prime's tale and the derogation of his pistol at that time is unknown. The feisty Colonel likely had some choice words.

Thus, we see that the Volcanic firearms were used in many ways, not only by private citizens, but by the U.S. military, individual officers and soldiers, civilian law enforcement agencies, explorers and western adventurers.

chapter nine

Volcanic Advertising and Promotion

NEW HAVEN ARMS CO.
Manufacturers of the Celebrated
Volcanic Repeating Fire Arms
COMPRISING
RIFLES, CARBINES AND PISTOLS,
WITH AMMUNITION WARRANTED WATER PROOF,
NEW HAVEN, CONN.
Depot for Sales,
267 BROADWAY, NEW YORK.
JOSEPH MERWIN, Agent.

Richard Labowskie collection

In the late 1870s, Oliver F. Winchester wrote of the firearm bearing his name that, "It has become a household word, and a household necessity on our western plains and mountains."[1] In the 1850s, however, "Volcanic" was hardly a household name, but Winchester was an astute businessman and a consummate entrepreneur who recognized the value of promotion and used several advertising media for the promotion of his firearms.

Broadsides and circulars were the most common advertising media in the immediate pre-Civil War period, and Winchester wisely used them to promote the Volcanic Repeating Arms Company. Newspapers also heralded the company's firearms, and O.F. Winchester exploited that medium through agents and friends, to get the "Volcanic" name before the public.[2] Other means were used as well including hand bills, pamphlets, magazines and city directories.

Reproduced here are several advertisements that all have features in common: they prominently display the company's name, and they all lavishly describe the attributes of the firearms and advertise their prices and that of the ammunition.

Daguerreotype of a .31 caliber Volcanic Pocket pistol from the estate of Oliver F. Winchester, probably taken and used for promotional purposes. This is the only known period photograph of a Volcanic firearm.

Donovan Williamson collection

CIRCULARS AND BROADSIDES

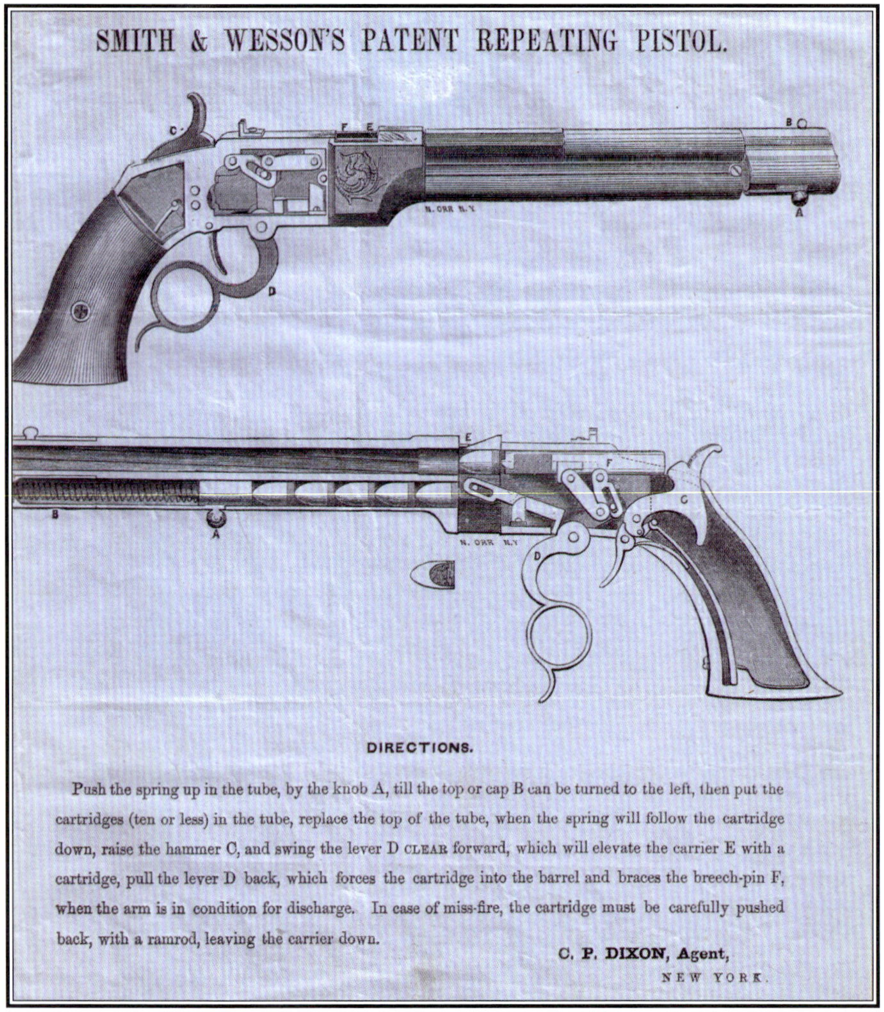
SMITH & WESSON'S PATENT REPEATING PISTOL.

DIRECTIONS.

Push the spring up in the tube, by the knob A, till the top or cap B can be turned to the left, then put the cartridges (ten or less) in the tube, replace the top of the tube, when the spring will follow the cartridge down, raise the hammer C, and swing the lever D CLEAR forward, which will elevate the carrier E with a cartridge, pull the lever D back, which forces the cartridge into the barrel and braces the breech-pin F, when the arm is in condition for discharge. In case of miss-fire, the cartridge must be carefully pushed back, with a ramrod, leaving the carrier down.

C. P. DIXON, Agent,
NEW YORK.

The earliest known, and likely unique handbill for the "SMITH & WESSON'S PATENT REPEATING PISTOL" was found by Roy Jinks, placed within the pages of a book titled, *The Life and Adventures of Kit Carson* by Dewitt C. Peters, published in 1858.[3] This circular, which measures 7¾ x 9¾ inches, is undated, circa 1853 to 1854 and lists C.P. Dixon as the New York agent. Earlier, Dixon was the agent for Robbins & Lawrence when they were marketing the Jennings rifle.[4] The handbill depicts a large-frame model with the top illustration showing the magazine spring retracted and the magazine block open. Note the engraving on the receiver. The bottom illustration was used, unchanged, on later broadsides of the Volcanic Repeating Arms Company. N. Orr was the artist. It did not list prices nor mention ammunition. Roy Jinks collection

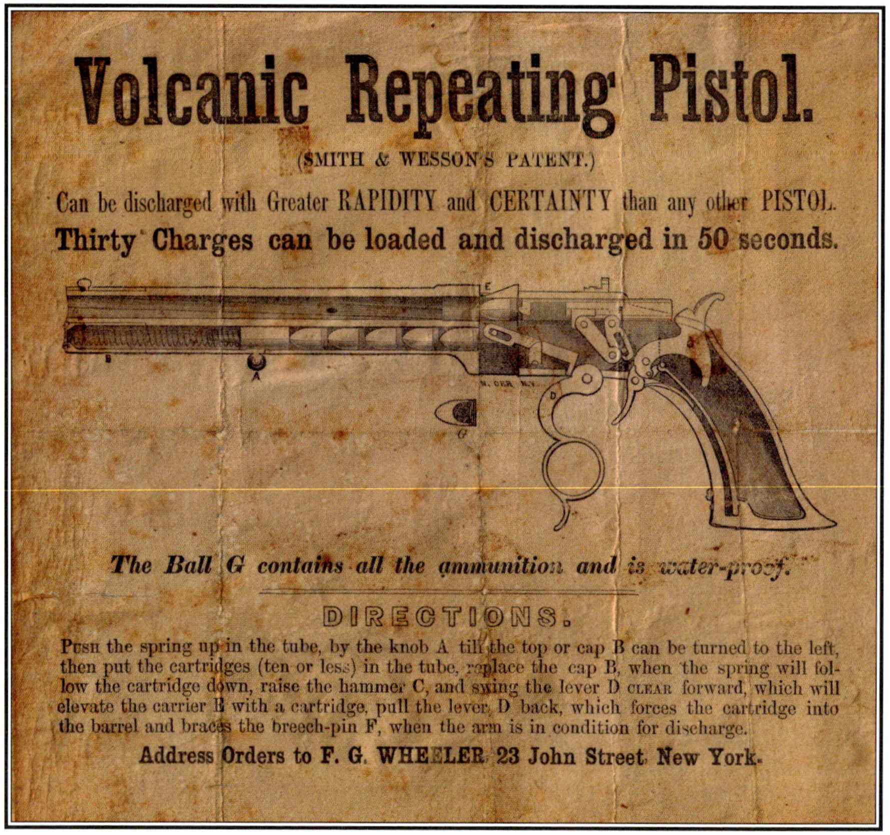
Volcanic Repeating Pistol.

(SMITH & WESSON'S PATENT.)

Can be discharged with Greater RAPIDITY and CERTAINTY than any other PISTOL.

Thirty Charges can be loaded and discharged in 50 seconds.

The Ball G contains all the ammunition and is water-proof.

DIRECTIONS.

Push the spring up in the tube, by the knob A till the top or cap B can be turned to the left, then put the cartridges (ten or less) in the tube, replace the cap B, when the spring will follow the cartridge down, raise the hammer C, and swing the lever D CLEAR forward, which will elevate the carrier E with a cartridge, pull the lever D back, which forces the cartridge into the barrel and braces the breech-pin F, when the arm is in condition for discharge.

Address Orders to F. G. WHEELER 23 John Street New York.

Undated, this is the earliest known circular (measuring 7¾ x 9¾ inches) advertising Volcanic firearms. Note that it is titled "Volcanic Repeating Pistol" rather than "Volcanic Repeating Fire-Arms" as on later circulars. Also note in parentheses below the title is (SMITH & WESSON'S PATENT.) This is the only time any of the volcanic advertising referred directly to Smith & Wesson, although later circulars indicated their 1854 patent date. F.G. Wheeler is the earliest known agent for the company. He was also an agent for the Waterville Manufacturing Company of Waterbury, Connecticut, a cutlery manufacturer who had his office at 23 John Street between 1853 and 1856. A number of New York gun makers and retailers had premises on John Street. Casey Nanz collection

Volcanic Repeating Fire-Arms,

[PATENTED 1854.]

Rifles, Carbines and Pistols loading with from 7 to 24 Balls.

Can be discharged with Greater RAPIDITY and Certainty than any other Pistol or Rifle.

Thirty Balls can be loaded and discharged in 50 seconds.

REPEATING FIRE ARMS.

AND PATENT LOADED BALL.

This Arm and Ammunition, lately introduced by the manufacturers into public notice, has deservedly met with great favor, its excellence is claimed to be far superior to ALL other Repeating Arms. It is simple in its construction, *compact* and *elegant* in its *proportions*, of great strength and durability, and NOT LIABLE to get out of order. The barrels are all RIFLED with great exactness. It can be loaded with seven to twenty-four Balls in four to ten seconds and can be discharged (in any weather) *twenty-five times* in *fifty seconds.* No cap—no priming—no recoil—no lateral discharge. The Ball is a *Patent Loaded Ball*, on the Minie principle, and is perfectly water-proof, and yet always SURE and SAFE.

The Ball G contains all the Ammunition and is Water-proof.

DIRECTIONS.

Push the spring up in the tube, by the knob A till the top or cap B can be turned to the left, then put the cartridges (ten or less) in the tube, replace the cap B, when the spring will follow the cartridge down, raise the hammer C, and swing the lever D CLEAR forward, which will elevate the carrier E with a cartridge, pull the lever D CLEAR back, which forces the cartridge into the barrel and braces the breech-pin F, when the arm is in condition for discharge. In case of mis-fire bring up another ball,

Manufacturers' Depot, 229 Broadway, cor. Barclay-st., New York.

J. W. POST, SOLE AGENT.

This early Volcanic Repeating Arms circular lists J.W. Post as agent, circa 1857. It is one of two circulars known that mentions "Volcanic Repeating Fire-Arms" and not the New Haven Arms Company. It advertises "Rifles, Carbines and Pistols loading with from 7 to 24 Balls" but does not price either the guns or ammunition as found on later broadsides. This circular measures 11 x 9 inches. Richard Labowskie collection

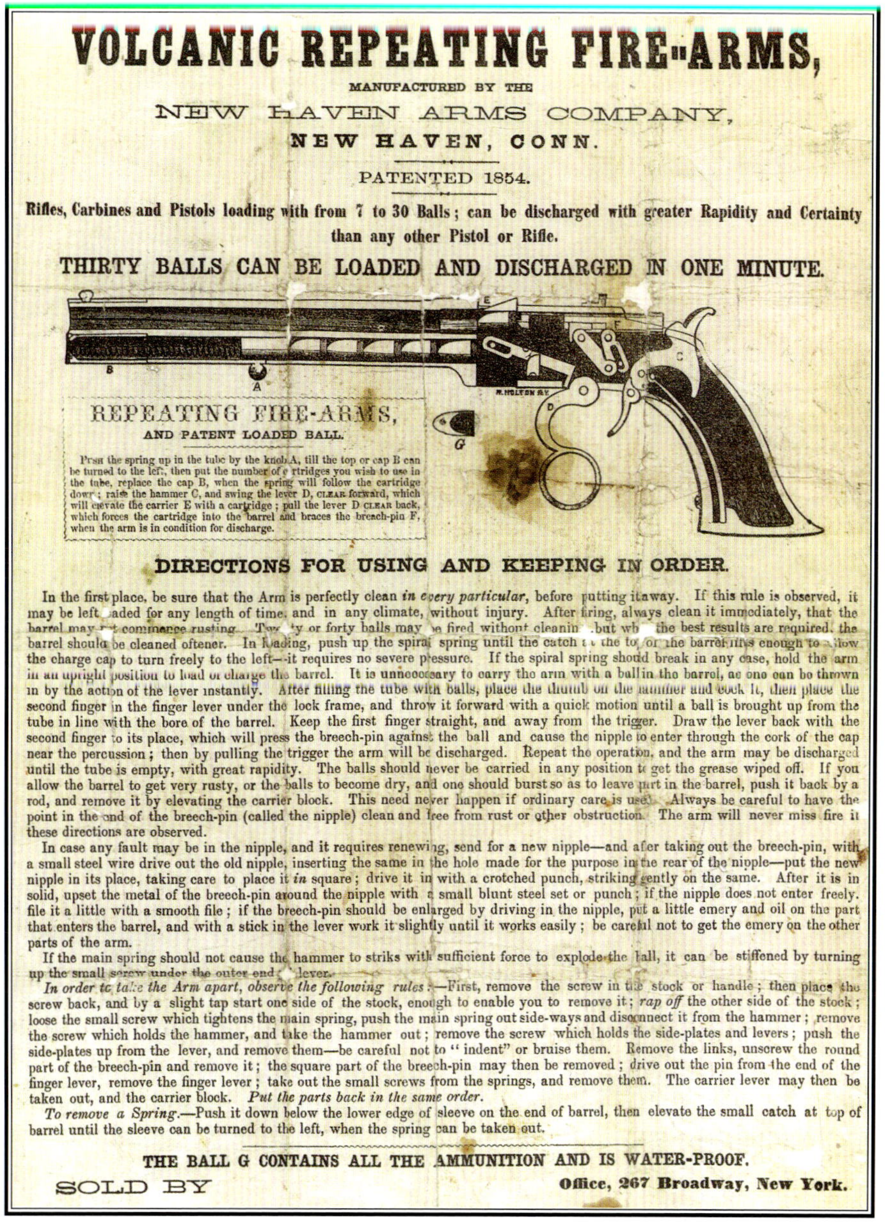

VOLCANIC REPEATING FIRE-ARMS,

MANUFACTURED BY THE

NEW HAVEN ARMS COMPANY,

NEW HAVEN, CONN.

PATENTED 1854.

Rifles, Carbines and Pistols loading with from 7 to 30 Balls; can be discharged with greater Rapidity and Certainty than any other Pistol or Rifle.

THIRTY BALLS CAN BE LOADED AND DISCHARGED IN ONE MINUTE.

REPEATING FIRE-ARMS,

AND PATENT LOADED BALL.

Push the spring up in the tube by the knob A, till the top or cap B can be turned to the left, then put the number of cartridges you wish to use in the tube, replace the cap B, when the spring will follow the cartridge down; raise the hammer C, and swing the lever D, CLEAR forward, which will elevate the carrier E with a cartridge; pull the lever D CLEAR back, which forces the cartridge into the barrel and braces the brench-pin F, when the arm is in condition for discharge.

DIRECTIONS FOR USING AND KEEPING IN ORDER.

In the first place, be sure that the Arm is perfectly clean *in every particular*, before putting it away. If this rule is observed, it may be left loaded for any length of time, and in any climate, without injury. After firing, always clean it immediately, that the barrel may not commence rusting. Twenty or forty balls may be fired without cleaning, but when the best results are required, the barrel should be cleaned oftener. In loading, push up the spiral spring until the catch at the top of the barrel lifts enough to allow the charge cap to turn freely to the left—it requires no severe pressure. If the spiral spring should break in any case, hold the arm in an upright position to load or charge the barrel. It is unnecessary to carry the arm with a ball in the barrel, as one can be thrown in by the action of the lever instantly. After filling the tube with balls, place the thumb on the hammer and cock it, then place the second finger in the finger lever under the lock frame, and throw it forward with a quick motion until a ball is brought up from the tube in line with the bore of the barrel. Keep the first finger straight, and away from the trigger. Draw the lever back with the second finger to its place, which will press the breech-pin against the ball and cause the nipple to enter through the cork of the cap near the percussion; then by pulling the trigger the arm will be discharged. Repeat the operation, and the arm may be discharged until the tube is empty, with great rapidity. The balls should never be carried in any position to get the grease wiped off. If you allow the barrel to get very rusty, or the balls to become dry, and one should burst so as to leave part in the barrel, push it back by a rod, and remove it by elevating the carrier block. This need never happen if ordinary care is used. Always be careful to have the point in the end of the breech-pin (called the nipple) clean and free from rust or other obstruction. The arm will never miss fire if these directions are observed.

In case any fault may be in the nipple, and it requires renewing, send for a new nipple—and after taking out the breech-pin, with a small steel wire drive out the old nipple, inserting the same in the hole made for the purpose in the rear of the nipple—put the new nipple in its place, taking care to place it *in* square; drive it in with a crotched punch, striking gently on the same. After it is in solid, upset the metal of the breech-pin around the nipple with a small blunt steel set or punch; if the nipple does not enter freely, file it a little with a smooth file; if the breech-pin should be enlarged by driving in the nipple, put a little emery and oil on the part that enters the barrel, and with a stick in the lever work it slightly until it works easily; be careful not to get the emery on the other parts of the arm.

If the main spring should not cause the hammer to striks with sufficient force to explode the ball, it can be stiffened by turning up the small screw under the outer end of lever.

In order to take the Arm apart, observe the following rules:—First, remove the screw in the stock or handle; then place the screw back, and by a slight tap start one side of the stock, enough to enable you to remove it; *rap off* the other side of the stock; loose the small screw which tightens the main spring, push the main spring out side-ways and disconnect it from the hammer; remove the screw which holds the hammer, and take the hammer out; remove the screw which holds the side-plates and levers; push the side-plates up from the lever, and remove them—be careful not to "indent" or bruise them. Remove the links, unscrew the round part of the breech-pin and remove it; the square part of the breech-pin may then be removed; drive out the pin from the end of the finger lever, remove the finger lever; take out the small screws from the springs, and remove them. The carrier lever may then be taken out, and the carrier block. *Put the parts back in the same order.*

To remove a Spring.—Push it down below the lower edge of sleeve on the end of barrel, then elevate the small catch at top of barrel until the sleeve can be turned to the left, when the spring can be taken out.

THE BALL G CONTAINS ALL THE AMMUNITION AND IS WATER-PROOF.

SOLD BY **Office, 267 Broadway, New York.**

This undated circular, titled "Volcanic Repeating Fire-Arms" makes no mention of the New Haven Arms Company. Joseph Merwin is prominently listed as the sole agent. Edmund lewis collection

A New Haven Arms Company circular with no agent listed, circa 1857–1858. It shows a cutaway view of a large-frame Smith & Wesson pistol rather than one of their own making. They also illustrate a Volcanic cartridge. The circular measures 9¼ x 12¼ inches. The artist was H. Holton. The illustration is the same as that found on the Smith & Wesson circular. Courtesy of R.L. Wilson[5]

J. HAPGOOD, a Boston gun store, lists itself on this circular, ca.1858, as agent for the sale of the VOLCANIC REPEATING ARMS. Many companies referred to Volcanic long arms as "rifles" rather than carbines. Casey Nanz collection

VOLCANIC
REPEATING FIRE-ARMS.

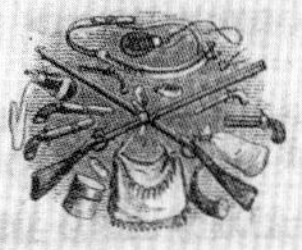

PATENTED 1854.

The New Haven Arms Company

Have recently obtained the entire control of the above Arms, and are now manufacturing them in the most perfect manner, and are prepared to furnish the trade on favorable terms.

The rapidity of execution of this Arm places it beyond all competition. The thirty shooter can be loaded and fired in less than one minute—a quickness and force of execution which is as much superior to the best revolvers as they are to the old muzzle loading single shooters.

The Ammunition is water-proof, hence it can be used in any weather, or loaded and hung up for months, or laid under water, and then fired with certainty.

Its safety from accidental discharge is a great consideration in its favor; for while the magazine (a tube running the whole length of the barrel) may be filled with balls, and thus the gun, in fact, be loaded from breech to muzzle, it is yet impossible, from any carelessness in handling, to discharge it. *Its construction* is simple and its workmanship most perfect, hence it is not easily got out of repair.

Its proportions are light, elegant, and compact, and the barrels are all rifled with great exactness. It requires no cap nor priming, no bullet mould nor powder flask. The powder and cap is contained in a loaded "minnie" ball of the best form and proportions, and is as sure as the best percussion caps.

It shoots with accuracy and greater force than any other arm can with double the powder used in this. Directions for use accompany each Arm. Balls are packed in tin cases, 200 each.

After this date, the prices will be as follows, viz.:

No. 1, 4 inch Pocket Pistol,	$12 00,	Plated and Engraved, $13 50,	Carrying	6 Balls.
No. 1, 6 " for Target Practice,	13 50,	" " " 15 00,	"	10 "
No. 2, 6 " Navy Pistol,	18 00,	" " " 20 00,	"	8 "
No. 2, 8 " " "	18 00,	" " " 20 00,	"	10 "
No. 2, 16 " Carbine,	30 00,	" " " 33 00,	"	20 "
No. 2, 20 " "	35 00,	" " " 38 00,	"	25 "
No. 2, 24 " "	40 00,	" " " 43 00,	"	30 "

AMMUNITION.

No. 1 Balls, 130 to the Pound, $10 per Thousand. No. 2 Balls, 66 to the Pound, $12 per Thousand.
(No. 1 Arms, require No. 1 Balls. No. 2 Arms, require No. 2 Balls.)

A liberal discount to the trade. Your orders are respectfully solicited.

NEW HAVEN ARMS COMPANY,
NEW HAVEN, CONN.

Office and Depot for Sales, 267 Broadway, New York.

NEW YORK, MAY 1, 1859. JOSEPH MERWIN, Agent.

Volcanic Repeating Fire Arms,

MANUFACTURED BY THE

NEW HAVEN ARMS COMPANY,

NEW HAVEN, CONN.

PATENTED 1854.

Rifles, Carbines and Pistols loading with from 7 to 30 Balls; can be discharged with greater Rapidity and Certainty than any other Pistol or Rifle.

THIRTY BALLS CAN BE LOADED AND DISCHARGED IN ONE MINUTE.

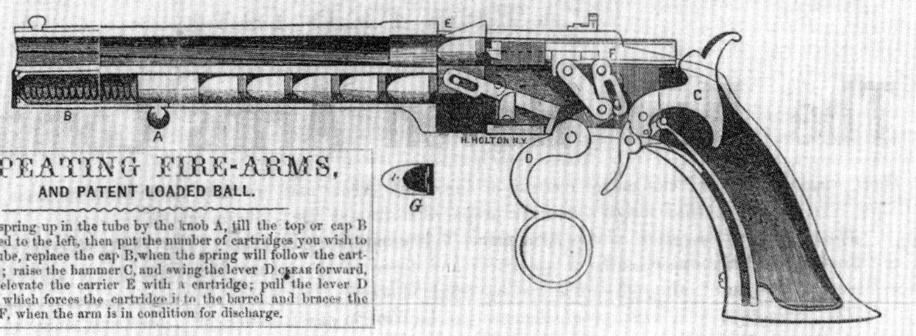

REPEATING FIRE-ARMS,
AND PATENT LOADED BALL.

PUSH the spring up in the tube by the knob A, till the top or cap B can be turned to the left, then put the number of cartridges you wish to use in the tube, replace the cap B, when the spring will follow the cartridge down; raise the hammer C, and swing the lever D CLEAR forward, which will elevate the carrier E with a cartridge; pull the lever D CLEAR back, which forces the cartridge into the barrel and braces the breech-pin F, when the arm is in condition for discharge.

DIRECTIONS FOR USING AND KEEPING IN ORDER.

In the first place, be sure that the Arm is perfectly clean *in every particular* before putting it away. If this rule is observed, it may be left loaded for any length of time, and in any climate, without injury. After firing always clean it immediately, that the barrel may not commence rusting. Twenty or forty balls may be fired without cleaning, but when the best results are required the barrel should be cleaned oftener. In loading, push up the spiral spring until the catch at the top of the barrel lifts enough to allow the charge cap to turn freely to the left—it requires no severe pressure. If the spiral spring should break in any case, hold the arm in an upright position to load or charge the barrel. It is unnecessary to carry the arm with a ball in the barrel, as one can be thrown in by the action of the lever instantly. After filling the tube with balls, place the thumb on the hammer and cock it; then place the second finger in the finger-lever under the lock frame, and throw it forward with a quick motion until a ball is brought up from the tube in line with the bore of the barrel. Keep the first finger straight, and away from the trigger. Draw the lever back with the second finger to its place, which will press the breech-pin against the ball and cause the nipple to enter through the cork of the cap near the percussion; then by pulling the trigger the arm will be discharged. Repeat the operation, and the arm may be discharged until the tube is empty with great rapidity. The balls should never be carried in any position to get the grease wiped off. If you allow the barrel to get very rusty, or the balls to become dry, and one should burst so as to leave part in the barrel, push it back by a rod, and remove it by elevating the carrier block. This need never happen if ordinary care is used. Always be careful to have the point in the end of the breech-pin (called the nipple) clean and free from rust or other obstruction. The arm will never miss fire if these directions are observed.

If the main spring should not cause the hammer to strike with sufficient force to explode the ball, it can be stiffened by turning up the small screw under the outer end of lever.

To remove a Spring.—Push it down below the lower edge of sleeve on the end of barrel, then elevate the small catch at top of barrel until the sleeve can be turned to the left, when the spring can be taken out.

THE BALL G CONTAINS ALL THE AMMUNITION AND IS WATER-PROOF.

Sold by JOSEPH MERWIN; Office, 267 Broadway, New York.

The front and back of a later New Haven Arms Company circular dated May 1, 1859, listing "JOSEPH MERWIN, Agent" with Office and Depot for Sales, 267 Broadway, New York. It lists the pistols and carbines available. No pistol-carbine is listed. The illustration is the same as that featured on the previous circular. The reverse of this circular was reprinted in the 1960s. The circular measures 9½ x 12 by inches. Original New Haven Arms Company circular. Richard Labowskie collection

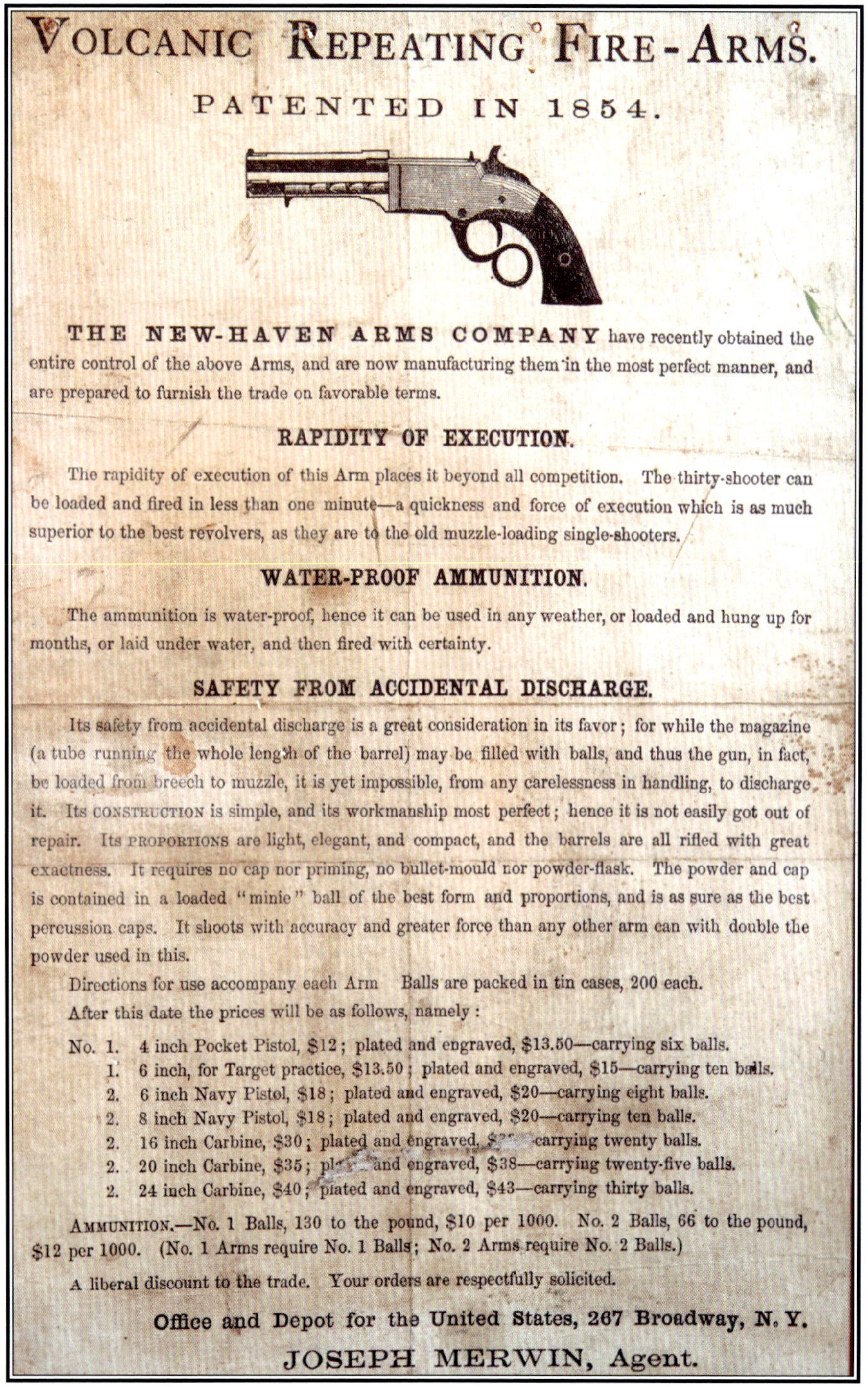

VOLCANIC REPEATING FIRE-ARMS.

PATENTED IN 1854.

THE NEW-HAVEN ARMS COMPANY have recently obtained the entire control of the above Arms, and are now manufacturing them in the most perfect manner, and are prepared to furnish the trade on favorable terms.

RAPIDITY OF EXECUTION.

The rapidity of execution of this Arm places it beyond all competition. The thirty-shooter can be loaded and fired in less than one minute—a quickness and force of execution which is as much superior to the best revolvers, as they are to the old muzzle-loading single-shooters.

WATER-PROOF AMMUNITION.

The ammunition is water-proof, hence it can be used in any weather, or loaded and hung up for months, or laid under water, and then fired with certainty.

SAFETY FROM ACCIDENTAL DISCHARGE.

Its safety from accidental discharge is a great consideration in its favor; for while the magazine (a tube running the whole length of the barrel) may be filled with balls, and thus the gun, in fact, be loaded from breech to muzzle, it is yet impossible, from any carelessness in handling, to discharge it. Its CONSTRUCTION is simple, and its workmanship most perfect; hence it is not easily got out of repair. Its PROPORTIONS are light, elegant, and compact, and the barrels are all rifled with great exactness. It requires no cap nor priming, no bullet-mould nor powder-flask. The powder and cap is contained in a loaded "minie" ball of the best form and proportions, and is as sure as the best percussion caps. It shoots with accuracy and greater force than any other arm can with double the powder used in this.

Directions for use accompany each Arm Balls are packed in tin cases, 200 each.

After this date the prices will be as follows, namely:

No. 1. 4 inch Pocket Pistol, $12; plated and engraved, $13.50—carrying six balls.
1. 6 inch, for Target practice, $13.50; plated and engraved, $15—carrying ten balls.
2. 6 inch Navy Pistol, $18; plated and engraved, $20—carrying eight balls.
2. 8 inch Navy Pistol, $18; plated and engraved, $20—carrying ten balls.
2. 16 inch Carbine, $30; plated and engraved, [illegible]—carrying twenty balls.
2. 20 inch Carbine, $35; pl[illegible] and engraved, $38—carrying twenty-five balls.
2. 24 inch Carbine, $40; plated and engraved, $43—carrying thirty balls.

AMMUNITION.—No. 1 Balls, 130 to the pound, $10 per 1000. No. 2 Balls, 66 to the pound, $12 per 1000. (No. 1 Arms require No. 1 Balls; No. 2 Arms require No. 2 Balls.)

A liberal discount to the trade. Your orders are respectfully solicited.

Office and Depot for the United States, 267 Broadway, N. Y.

JOSEPH MERWIN, Agent.

Edmund Lewis collection

Left: This is the largest New Haven Arms Company broadside known. Measuring 11 x 16 inches, it was printed in 1857, shortly after the failure of the Volcanic Repeating Arms Company. Note the first line, which states that, "THE NEW HAVEN ARMS COMPANY have recently obtained the entire control of the above Arms and are now manufacturing them in the most perfect manner…" This same advertisement appeared in *The Knickerbocker: or, New-York Monthly Magazine*, in 1859. It is the first time a New Haven Arms Company product is illustrated. The illustration is of the 3½-inch Pocket pistol, although it is designated and advertised as a "four inch" pistol by the company. The broadside also announces that the company "have recently obtained the entire control of the above Arms." The claims of "RAPIDITY OF EXECUTION, WATER-PROOF AMMUNITION AND SAFETY FROM ACCIDENTAL DISCHARGE" are more prominently noted than on the earlier circulars. JOSEPH MERWIN remained the agent. Edmund Lewis collection

NEWSPAPERS

Perhaps the most famous advertisement for the Volcanic repeating rifle was a piece published in *Frank Leslie's Illustrated Newspaper* of October 9, 1858, that described the Volcanic with great enthusiasm:

"It combines every requisite in such a weapon, with many advantages which no similar invention has yet succeeded in attaining. It is placed beyond all competition by the rapidity of its execution. Thirty shots can be fired in less than one minute-a really marvelous rapidity, in which it far outdoes the best revolving firearms yet produced. Its ammunition has the advantage of compactness, lightness, and of being water-proof. As will be seen in our diagram, the entire charge consists in a bullet of the Minie pattern, in which both charge and priming are contained, and of which sixty weigh only one pound. What an improvement upon the heavy cartridge or powder-flask that it has hitherto been necessary to carry! The balls may be soaked in water with perfect impunity, and can be kept any length of time in any

FRANK LESLIE'S ILLUSTRATED NEWSPAPER

No. 149.—VOL. VI.] NEW YORK, SATURDAY, OCTOBER 9, 1858. [PRICE 6 CENTS.

climate, without losing their explosive force; nor can they be exploded by contact with flame.

One of the principal recommendations of the Volcanic repeating rifle is its safety from accidental discharge, as, while the magazine (a tube running the whole length of the barrel), may be filled with balls, and thus the gun, in fact, be loaded from breech to muzzle, it is yet impossible from any carelessness in handling, to discharge it. Its construction is simple and its workmanship most perfect, hence it is not easily got out of repair.

The manufacture of these firearms-of which several sizes, as well pistols as rifles, are produced-was commenced in 1855, and is now carried on by the New Haven Arms Company, of New Haven, Conn., where a large factory is established, employing, on an average, some fifty hands. The depot of the company, a very handsome store, is at No. 267 Broadway, New York."[6]

With publicity like this, it would seem that the New Haven firearms would have been in greater demand, but those requiring a more dependable arm continued to use the more reliable muzzleloaders and breechloaders.

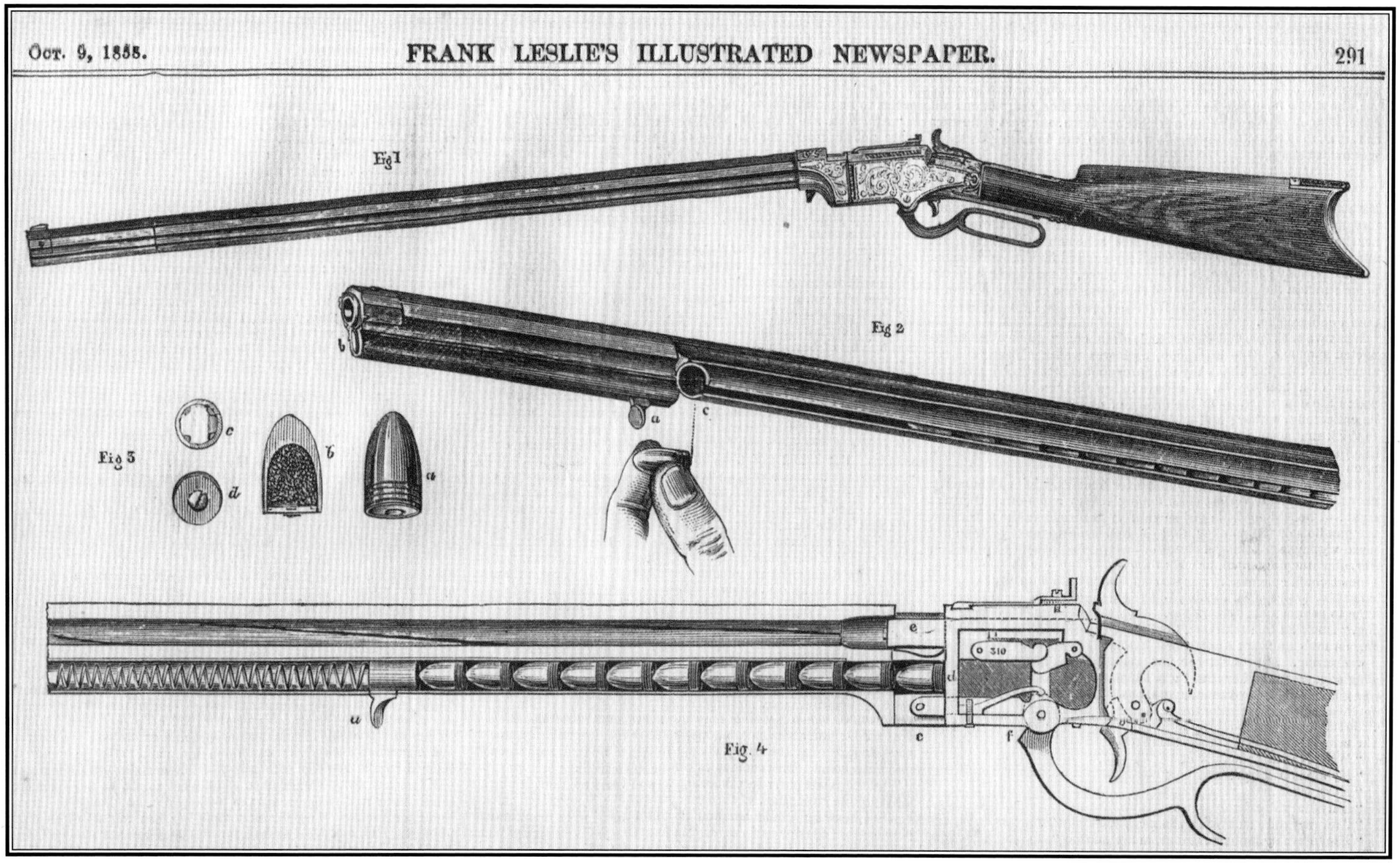

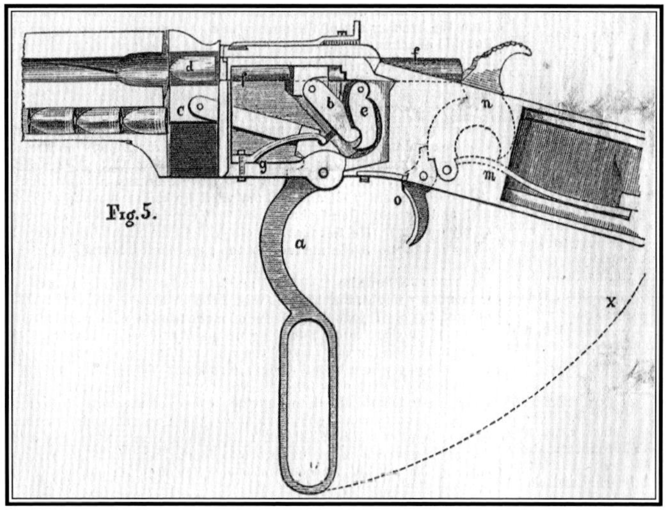

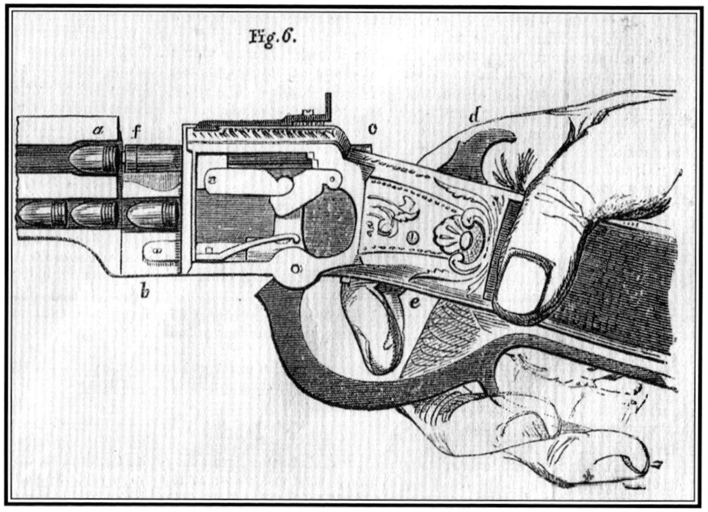

From Frank Leslie's *Illustrated Newspaper* dated October 9, 1858.
Edmund Lewis collection

Many newspapers in the East and Midwest trumpeted the virtues of the Volcanic pistol along with articles that appeared in the *Scientific American* and other journals. These are several examples courtesy of Herbert G. Houze:

"Volcanic Repeating Pistol
THE SUBSCRIBERS HAVE JUST
Received a small lot of the justly
celebrated Volcanic Repeating Pistols.
Thirty cartridges can be loaded and
discharged in 60 seconds. The ammunition
is Water-Proof, superior to anything yet out.
H. J. NAZRO & CO."
Milwaukee Daily Sentinel
May 26, 1855

Scientific American.
THE ADVOCATE OF INDUSTRY, AND JOURNAL OF SCIENTIFIC, MECHANICAL, AND OTHER IMPROVEMENTS
VOLUME XII. NEW-YORK, NOVEMBER 29, 1856. NUMBER 12.

An officer in the British Army had these kind, but somewhat incredulous, comments about the Volcanic pistol, which were published in the *Scientific American* of November 29, 1856:

"Col. Hay, of the British army, recently tried his hand with the 'Volcanic Repeating Pistol,' a Yankee invention. The pistol used on the occasion was an eight inch barrel, which discharges nine balls in rapid succession. The Colonel fired the arm twenty-seven times, making a number of shots which would do credit to a rifleman. He first fired at an eight-inch diameter target at 100 yards, putting nine balls inside the ring. He then moved back to a distance of 200 yards, and fired nine balls more, hitting the mark seven times. He then moved back 100 yards further, a distance of 300 yards from the mark, and placed five of the nine balls inside the ring, and hitting the 'bulls eye' twice. The man who beats that may brag."[7]
Scientific American,
Vol. 12, Issue 12, November 29, 1856

A PROCLAMATION!

"...One thousand men with the VOLCANIC REPEATING RIFLE, which carries thirty balls and can be loaded and discharged thirty times in less than one minute. One thousand men, armed with this Rifle would be equal to thirty thousand men. Armed with any other weapon..."
The New York Times, February 5, 1859

The hyperbole is stunning!

A Proclamation!
THIRTY MILLION DOLLARS SAVED TO THE COUNTRY.
THE PRESIDENT NO LONGER THINKS OF BUYING CUBA.

Argument and entreaty are exhausted. He has now come to a very sensible conclusion, viz.: to arm one thousand men with the VOLCANIC REPEATING RIFLE, which carries thirty balls, and can be loaded and discharged thirty times in less than one minute. One thousand men, armed with this Rifle. would be equal to thirty thousand armed with any other weapon. If our worthy *Chief Magistrate* does not bestir himself the Cubans will be ahead of him, as they are now buying them in large quaatities, and will soon be able to defend themselves against any force that can be brought against them. The balls are all waterproof—hence their great superiority over any other arm. No changes of climate, or even immersion in water, affects them.

Colt's Pistol, so long considered the *neplus ultra* of Firearms, is now fairly outdone by this new invention. Their beauty of construction and simplicity of movement are the wonder and admiration of all who have seen them. As was remarked a few days since, by one of our most accomplished commanders, "The man who invented that gun was a benefactor to his race, and the nation that was in possession of such a weapon need never fear even the combined armies of the world, if their cause was just."

Guns and Pistols, carrying from seven to thirty balls. This arm is now on exhibition and for sale, at wholesale and retail, at the Company's dépôt, No. 267 Broadway, New-York. JOSEPH MERWIN,
Sole Agent for United States.

Also agent for the celebrated G. D., C. & T., and Waterproof Percussion Caps.

!!EXCELSIOR!!

The prolific agent Joseph Merwin placed this notice in *The New York Times* on August 31, 1859, declaring that both *"The MILITARY and the PUBLIC are respectfully invited to call and examine this TRIUMPH OF AMERICAN INGENUITY..."* The military expressed very limited interest in the company's firearms.

HO! FOR PIKE'S PEAK!

At the time when men found themselves in need of reliable firearms, Westward expansion doubtless led Joseph Merwin, the ever-consummate advertising man, to place this notice in *The New York Times* on March 12, 1859, appealing to the mothers of those striking out for the newly discovered gold diggings in Colorado.

Ho! for Pike's Peak!
AND THE GOLD DIGGINGS!

No mother's son should be allowed to go to the Gold Diggings without one of those VOLCANIC-REPEATING RIFLES or PISTOLS, which carry from seven to thirty balls.

Without doubt, they are the most complete weapon ever introduced with the water-proof ball, which is not effected by any changes of climate, or subject to the many mishaps to which the ordinary revolver is liable to.

With a good conscience and a Volcanic Pistol, you may sleep soundly and safely, with your hard-earned nuggets by your side.

My friend, if you have regard for your life or your scalp, do not fail to get one or more of these invaluable weapons before starting on your journey. Buy for yourself, and advise your friends to do the same. A word to the wise is sufficient.

Rifles, Carbines and Pistols, carrying from 7 to 30 balls, twenty-five per cent. less than Colt's and so much better.
JOSEPH MERWIN, Agent,
No. 267 Broadway, New-York.

Also, Sole Agent for the American G. D. and Waterproof Percussion Caps.

!! EXCELSIOR !!
THE VOLCANIC REPEATING FIREARMS,
RIFLES, CARBINES AND PISTOLS,

Loading with from seven to thirty balls, can be discharged with greater rapidity and certainty than any other pistol or rifle made in the world.

THIRTY BALLS CAN BE LOADED
AND DISCHARGED IN ONE MINUTE,

and, on account of the great precision of manufacture, the most unerring execution can be done.

The frequent necessity and calls to use a reliable SHOOTING-ARM, either for pleasure or self-defence, renders it important that all should be provided in case of need.

The MILITARY and the PUBLIC are respectfully invited to call and examine this TRIUMPH OF AMERICAN INGENUITY, at the New-Haven Arms Company Dépôt, or address
JOSEPH MERWIN, Agent, No. 267 Broadway.

CITY DIRECTORIES AND ALMANACS

Almanacs and city directories usually carried business advertisements of many different types, including those for firearms. Following are three examples:

The Tribune Almanac, New York, 1857. (Reproduced at right) J.W. Post is listed as the sole agent. The products are described as a "6 inch barrel Pistol", an "8 inch barrel Pistol", a "16½ inch barrel Rifle Carbine" and a "16½ barrel Detached Breech Rifle Carbine". It is again apparent that the company itself listed their long arms as both rifles and carbines.

128 GUNS AND PISTOLS.

The great King of kings
Hath in the table of his law commanded
That thou shalt do no murder; wilt thou then
Spurn at his edict and fulfil a man's.—*Shakspeure.*

PROV.—Have not thy cloak to be made when it begins to rain.

JOSEPH MERWIN, AGENT,

VOLCANIC REPEATING FIRE-ARMS,

269 BROADWAY, N. Y.

THE RAPIDITY of execution of this Arm places it beyond all competition. The thirty-shooter can be loaded and fired in less than one minute—a quickness and force of execution which is as much superior to the best revolvers as they are to the old muzzle-loading single-shooters.

ITS SAFETY from accidental discharge is a great consideration in its favor; for while the magazine (a tube running the whole length of the barrel) may be filled with balls, and thus the gun, in fact, be loaded from breech to muzzle, it is yet impossible from any carelessness in handling to discharge it. Its construction is simple and its workmanship most perfect, hence it is not easily got out of repair.

Its proportions are light, elegant and compact, and the barrels are all rifled with great exactness. *It requires no cap nor priming, no bullet-mould or powder-flask.* The powder and cap is contained in a loaded Minnie ball, of the best form and proportions, and is as sure as the best percussion caps.

Its shoots with accuracy and greater force than any other arm can with double the powder used in this. Directions for use accompany each Arm. Balls are packed in tin cases, 200 each.

NOTE.—*The above arms are certainly one of the wonders of the age. Mr. Merwin is always at his post and will be happy to serve all.*—EDITOR.

Edmund Lewis collection

Edmund Lewis collection

VOLCANIC REPEATING FIRE-ARMS. 77

[PATENTED 1854.]

RIFLES, CARBINES AND PISTOLS LOADING WITH FROM 7 TO 24 BALLS!

Can be discharged with Greater RAPIDITY and Certainty than any other Pistol or Rifle.

Thirty Balls can be loaded and discharged in 50 seconds.

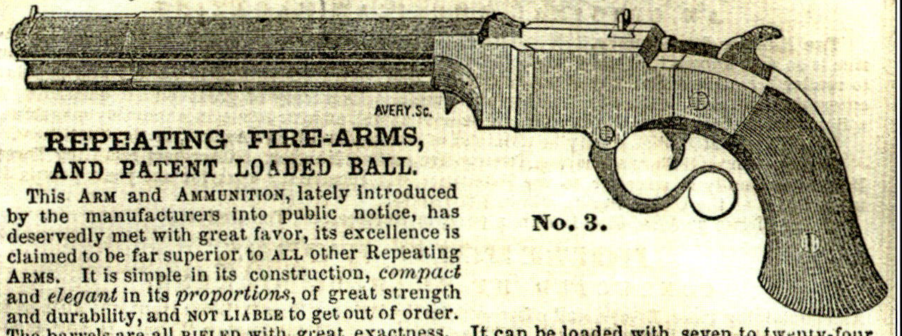

REPEATING FIRE-ARMS, AND PATENT LOADED BALL.

No. 3.

This ARM and AMMUNITION, lately introduced by the manufacturers into public notice, has deservedly met with great favor, its excellence is claimed to be far superior to ALL other Repeating ARMS. It is simple in its construction, *compact* and *elegant* in its *proportions*, of great strength and durability, and NOT LIABLE to get out of order. The barrels are all RIFLED with great exactness. It can be loaded with seven to twenty-four balls in four to ten seconds, and can be discharged (in any weather) *twenty-five times* in *fifty seconds.* No cap—no priming—no recoil—no lateral discharge. The Ball is a *Patent Loaded Ball*, on the Minie principle, and is perfectly water-proof, and yet always SURE and SAFE, and of unequalled range and accuracy.

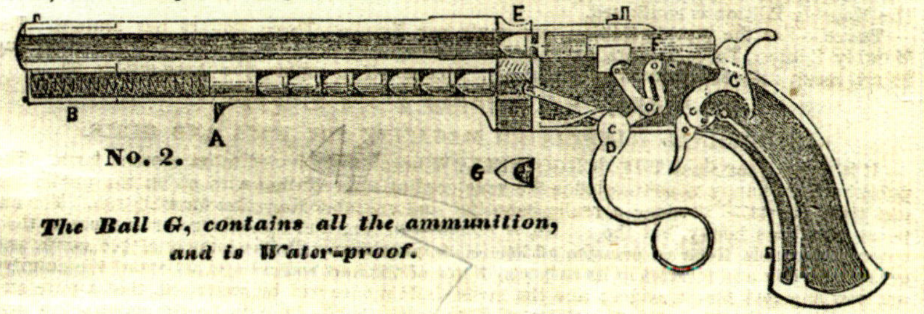

No. 2.

The Ball G, contains all the ammunition, and is Water-proof.

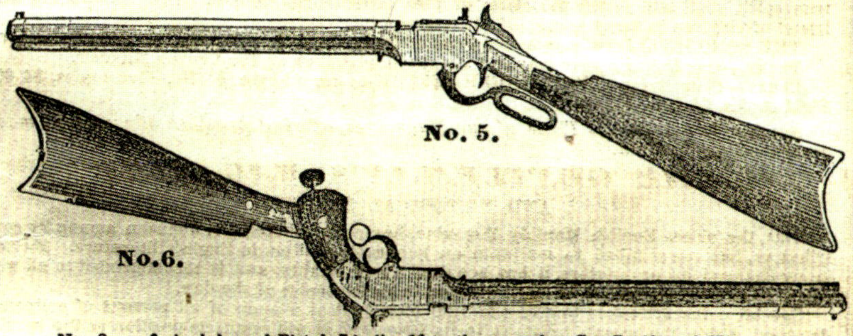

No. 5.

No.6.

No. 2 or 6 inch barrel Pistol, 7 balls, 66 to the pound.—*See Sectional View.*
No. 3 or 8 inch barrel Pistol, 9 balls, 66 to the pound.
No. 5 or 16½ inch barrel Rifle Carbine, 20 balls, 66 to the pound.
No. 6 or 16½ inch barrel Detached Breech Rifle Carbine, 20 balls, 66 to the pound.

DIRECTIONS.—*See No. 2.*

Push the spring up in the tube by the knob A, till the top or cap B can be turned to the left, then put the cartridges (ten or less) in the tube, replace the cap B, when the spring will follow the cartridge down, raise the hammer C, and swing the lever D CLEAR forward, which will elevate the carrier E with a cartridge, pull the lever D clear back, which forces the cartridge into the barrel and braces the breech-pin F, when the arm is in condition for discharge. In case of mis-fire bring up another ball.

Circulars with prices at wholesale and retail, can be had by addressing J. W. POST, 229 *Broadway, N. Y.*, Sole Agent for the sale of the Volcanic Repeating Fire-Arms for the U. States.

Manufacturers' Depot, 229 Broadway, corner Barclay St., New York.

J. W. POST, Sole Agent,

Joseph Merwin, listed as agent on several circulars, was the New York City agent for the New Haven Arms Company from 1858 to 1859.[4] In following years, other firearms dealers and agents were established in Philadelphia, Boston, New Orleans and San Francisco.[5]

An advertisement in Carroll's New York City Directory (reproduced on page 123) that is undated but is later in the period of production since Joseph Merwin is listed as the New York agent. It has information not shown on earlier broadsides and boasts: *"Its proportions are light, elegant and compact, and the barrels are all rifled with great exactness. It requires no cap nor priming, no bullet-mould or powder-flask. The powder and cap is contained in a loaded Minnie ball, of the best form and proportions, and is as sure as the best percussion caps."*

• • •

This Volcanic broadside of October 1859 (at right) includes the first use of testimonials. The testimonials are printed at the bottom of the broadside. One, from an ex-captain of the Clipper Ship *Staghound* states, *"I consider the Volcanic Repeating Pistol the* ne plus ultra *of Repeating or Revolving Arms, and far superior in many respects to Colt's much extolled Revolver. I have fired, myself, over 200 shots from it without even wiping the barrel — this is an advantage which no other Arm I know possesses. I have had the pistol with me at sea for more than eighteen months, on a voyage around the world, and find…it will keep free from rust far more so than Colt's… I have shown the Pistol to my friends in San Francisco, Hong Kong, Manila, Canton and Shanghaie,* [sic] *and they were much pleased with it."*

• • •

The other testimonial from the Commander of the ship *Star of the Union*, states, *"I have used a Volcanic Repeating Pistol for some months, on my last voyage to San Francisco, and in all that constitutes a good Pistol or Fire Arm, it has no equal, and excels all others I have ever seen in rapidity, efficiency and certainty of execution. Its peculiar merit for sea service is in the nature of the Ball, which containing the Ammunition, is water-proof, and cannot be damaged by any change of climate, but is sure fire even after having been loaded for months."* At this late date, it is indicated that, "All orders may be addressed to NEW HAVEN ARMS COMPANY, NEW HAVEN, CONN." No agent is listed.

In summary, the Smith & Wesson Company, the Volcanic Repeating Arms Company and the New Haven Arms Company placed their products before the public, through a variety of advertising media as illustrated. Although advertisements often included exaggerated performance claims, they always included interesting and detailed information not always provided by other firearms manufacturers of the period.

VOLCANIC
REPEATING FIRE ARMS,
MANUFACTURED BY THE
NEW HAVEN ARMS COMPANY,
NEW HAVEN, CONN.
(PATENTED, 1854.)

The above named Company having purchased all the Patent rights on this Arm and its ammunition, (some eight or ten in number,) the inventions of as many of the most ingenious mechanics of the country, who have spent years in bringing this wonderful triumph of genius to perfection, are now prepared to manufacture them in a perfect manner, and offer them for sale as the most powerful and effective weapon of defense ever invented. They are made of all sizes, from a four inch Pocket Pistol, carrying six balls, to a twenty-four inch Rifle, carrying thirty balls.

The rapidity of execution of this Arm places it beyond all competition. The thirty shooter can be loaded and fired in less than one minute—a quickness and force of execution which is as much superior to the best revolvers, as they are to the old muzzle loading single shooters.

The Ammunition is water-proof, hence it can be used in any weather, or loaded and hung up for months, or laid under water, and then fired with certainty.

Its safety from accidental discharge is a great consideration in its favor; for while the magazine (a tube running the whole length of the barrel) may be filled with balls, and thus the gun, in fact, be loaded from breech to muzzle, it is yet impossible, from any carelessness in handling, to discharge it. *Its construction* is simple and its workmanship most perfect, hence it is not easily got out of repair.

Its proportions are light, elegant and compact, and the barrels are all rifled with great exactness. It requires no cap nor priming, no bullet mould nor powder flask. The powder and cap is contained in a loaded "minnie" ball of the best form and proportions, and is as sure as the best percussion caps.

It shoots with accuracy and *greater force* than any other Arm can with double the powder used in this. Directions for use accompany each Arm. Balls are packed in tin cases, 200 each.

LIST OF MANUFACTURERS' PRICES.

No. 1,	4 inch	Pocket Pistol,	$12.00,	Plated and Engraved,	$13.50,	Carrying	6	Balls.
" 1,	6 "	for Target Practice,	13.50,	" "	15.00,	"	10	"
" 2,	6 "	Navy Pistol,	18.00,	" "	20.00,	"	8	"
" 2,	8 "	" "	18.00,	" "	20.00,	"	10	"
" 2,	16 "	Carbine,	30.00,	" "	33.00,	"	20	"
" 2,	20 "	"	35.00,	" "	38.00,	"	25	"
" 2,	24 "	"	40.00,	" "	43.00,	"	30	"

Plating and Engraving, from $2.50 to $5.00 extra, per Arm.

AMMUNITION.

No. 1 Balls, 130 to the Pound, $10 per M. No. 2 Balls, 66 to the Pound, $12 per M. (No. 1 Arms, require No. 1 Balls. No. 2 Arms, require No. 2 Balls.)

The numbers 1 and 2 designate the size of the bore, and the Balls are numbered to correspond.

A liberal discount to the trade.

We select the following from numerous testimonials, as the service to which the Arms were subjected was most severe, from the rapid action of salt water upon all metals.

New York, March 10th, 1855.

Gent.:—I consider the Volcanic Repeating Pistol the *ne plus ultra* of Repeating or Revolving Arms, and far superior in many respects to Colt's much extolled Revolver. I have fired, myself, over 200 shots from it without even wiping the barrel—this is an advantage which no other Arm I know of possesses. I have had the Pistol with me at sea for more than eighteen months, on a voyage around the world, and find that, with the most common care, it will keep free from rust, far more so than Colt's. I find the Balls as good now as when I left New York. I have shown the Pistol to my friends in San Francisco, Hong Kong, Manilla, Canton and Shanghaie, and they were much pleased with it.

C. F. W. BEHM, Late of Clipper Ship Stag Hound.

New York, 23d November, 1856.

Gent.:—I have used a Volcanic Repeating Pistol for some months, on my last voyage to San Francisco, and in all that constitutes a good Pistol or Fire Arm, it has no equal, and excels all others I have ever seen in rapidity, efficiency and certainty of execution. Its peculiar merit for sea service is in the nature of the Ball, which containing the Ammunition, is water-proof, and cannot be damaged by any change of climate, but is sure fire even after having been loaded for months.

Signed, FRED'K A. STALL, Commander Ship Star of the Union.

All orders may be addressed to

NEW HAVEN ARMS COMPANY,
New Haven, Conn.

October, 1859.

T. J. STAFFORD, PRINTER, 88 STATE STREET, (STAFFORD BUILDING.)

Edmund Lewis collection

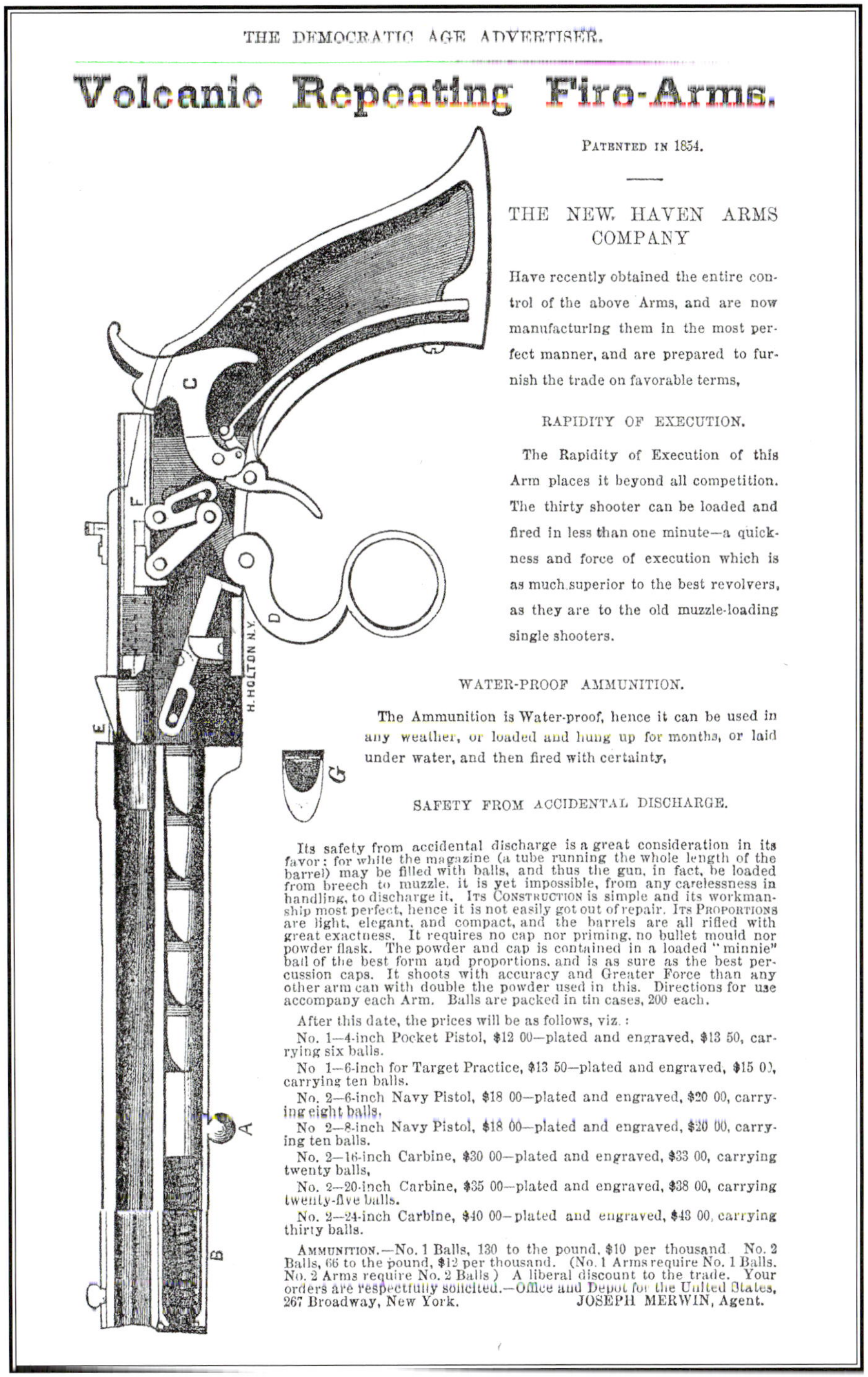

THE DEMOCRATIC AGE ADVERTISER.

Volcanic Repeating Fire-Arms.

PATENTED IN 1854.

H. HOLTON N.Y.

THE NEW HAVEN ARMS COMPANY

Have recently obtained the entire control of the above Arms, and are now manufacturing them in the most perfect manner, and are prepared to furnish the trade on favorable terms,

RAPIDITY OF EXECUTION.

The Rapidity of Execution of this Arm places it beyond all competition. The thirty shooter can be loaded and fired in less than one minute—a quickness and force of execution which is as much superior to the best revolvers, as they are to the old muzzle-loading single shooters.

WATER-PROOF AMMUNITION.

The Ammunition is Water-proof, hence it can be used in any weather, or loaded and hung up for months, or laid under water, and then fired with certainty,

SAFETY FROM ACCIDENTAL DISCHARGE.

Its safety from accidental discharge is a great consideration in its favor; for while the magazine (a tube running the whole length of the barrel) may be filled with balls, and thus the gun, in fact, be loaded from breech to muzzle, it is yet impossible, from any carelessness in handling, to discharge it. ITS CONSTRUCTION is simple and its workmanship most perfect, hence it is not easily got out of repair. ITS PROPORTIONS are light, elegant, and compact, and the barrels are all rifled with great exactness. It requires no cap nor priming, no bullet mould nor powder flask. The powder and cap is contained in a loaded "minnie" ball of the best form and proportions, and is as sure as the best percussion caps. It shoots with accuracy and Greater Force than any other arm can with double the powder used in this. Directions for use accompany each Arm. Balls are packed in tin cases, 200 each.

After this date, the prices will be as follows, viz.:

No. 1—4-inch Pocket Pistol, $12 00—plated and engraved, $13 50, carrying six balls.

No 1—6-inch for Target Practice, $13 50—plated and engraved, $15 00, carrying ten balls.

No. 2—6-inch Navy Pistol, $18 00—plated and engraved, $20 00, carrying eight balls,

No 2—8-inch Navy Pistol, $18 00—plated and engraved, $20 00, carrying ten balls.

No. 2—16-inch Carbine, $30 00—plated and engraved, $33 00, carrying twenty balls,

No. 2—20-inch Carbine, $35 00—plated and engraved, $38 00, carrying twenty-five balls.

No. 2—24-inch Carbine, $40 00—plated and engraved, $43 00, carrying thirty balls.

AMMUNITION.—No. 1 Balls, 130 to the pound, $10 per thousand. No. 2 Balls, 66 to the pound, $12 per thousand. (No. 1 Arms require No. 1 Balls. No. 2 Arms require No. 2 Balls.) A liberal discount to the trade. Your orders are respectfully solicited.—Office and Depot for the United States, 267 Broadway, New York. JOSEPH MERWIN, Agent.

Left. *From* The Democratic Age Advertiser, *February 1859, depicting an early large-frame Smith & Wesson pistol with the early magazine follower. The artist is H. Holton. It measures 6 x 9 inches.*

Edmund Lewis collection

Bruce Jordan collection

Small-frame Smith & Wesson pistol, Type I, serial number 13A.
Bruce Jordan collection

chapter ten

The Henry Rifle and Its Successors

THE NEW HAVEN ARMS COMPANY & BENJAMIN TYLER HENRY

With the New Haven Arms Company facing insolvency, the company president, Oliver F. Winchester, directed B. Tyler Henry, the shop superintendent, to develop a self-contained metallic cartridge and to redesign the Volcanic carbine to accept such a cartridge. The key to the success of the Henry rifle would be the development of the Henry cartridge. Without it, the Henry rifle would never have been any more successful than the Volcanic.[1] Winchester convinced the board of directors that the future of the company had to be placed in Henry's hands.[2] Afterward, the firm dropped the Volcanic line and put all its energy into the new Henry rifle chambered in .44 rimfire. The Henry cartridge became the standard ammunition for the "Improved Volcanic," as the Henry Rifle was first called. It was later called the Model 1866 Winchester. All design characteristics of the Smith & Wesson, Volcanic and New Haven barrel assemblies were to be retained in the new model rifle. Only the size was increased. The breech bolt and firing pin were redesigned for better ignition, and an extractor made of spring steel was fitted to the breech bolt.

Production, however, of the Henry rifle was delayed until sufficient funds became available. One effort made to generate money in 1860 involved the firm's agreement to manufacture 3,000 percussion revolvers based upon John Walch's U.S. Patent No. 22,905, issued on February 8, 1859, for the Walch Fire-Arms Company of New York. While it was anticipated that this contract would generate $8,000 in profit from the total contract price of $26,000, the Walch Fire-Arms Company failed prior to the completion of the contract. The New Haven Arms Company, after seeking legal recourse, received only the revolvers as compensation.[3]

Henry completed his work, and on October 16, 1860, was granted U.S. Patent No. 30,446, entitled "MAGAZINE FIRE ARM", claiming only his improvements in a magazine firearm. He did not claim the self-contained metallic cartridge as his invention. He knew well that such cartridges had been previously invented and patented, but he did include a drawing of his cartridge in the patent with the following description: "FIGURE 12 REPRESENTS THE AMMUNITION USED — A CONICAL OR ROUND BALL INSERTED INTO A METAL CARTRIDGE, HAVING A HOLLOW ROUND FLANGE AROUND THE BASE CONTAINING A PRIMING OF FULMINATE POWDER". Thus was born the famous Henry cartridge, more than six years after the invention of the toggle-link lever action tubular-magazine repeater.[4]

Benjamin Tyler Henry, 1821–1898

The Henry cartridge became the standard ammunition for the Henry rifle and for the improved Henry, better

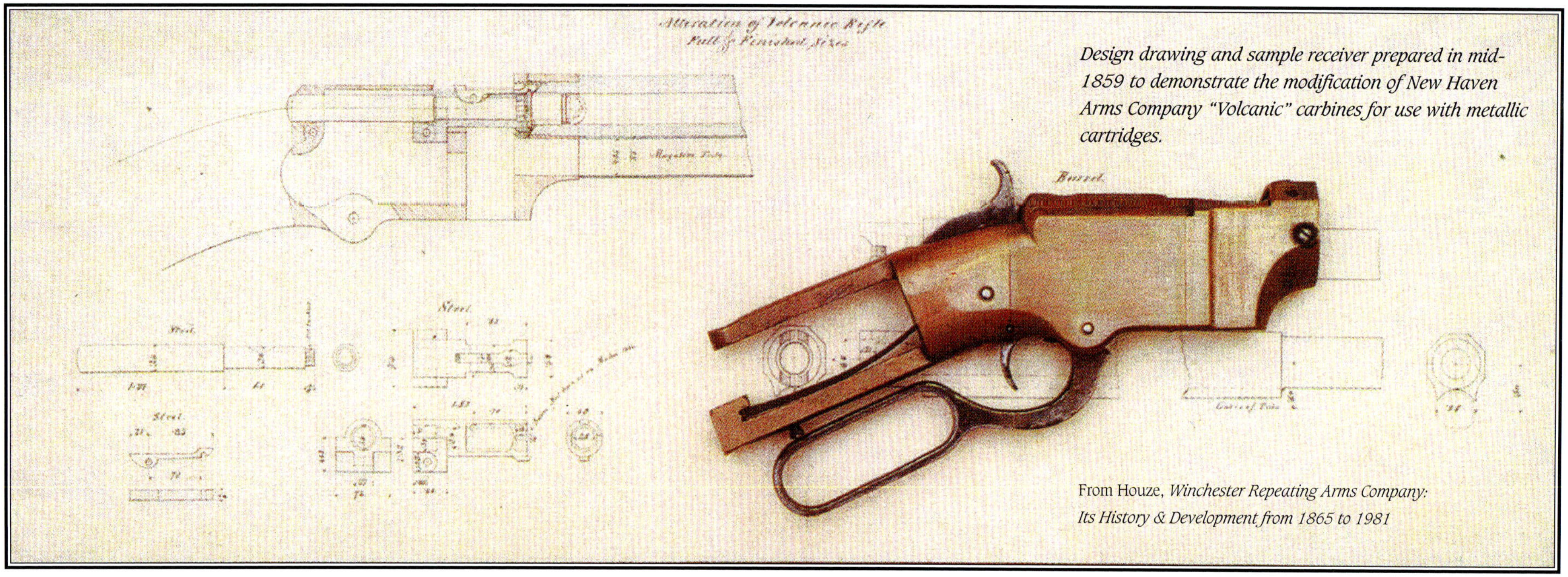

Design drawing and sample receiver prepared in mid-1859 to demonstrate the modification of New Haven Arms Company "Volcanic" carbines for use with metallic cartridges.

From Houze, *Winchester Repeating Arms Company: Its History & Development from 1865 to 1981*

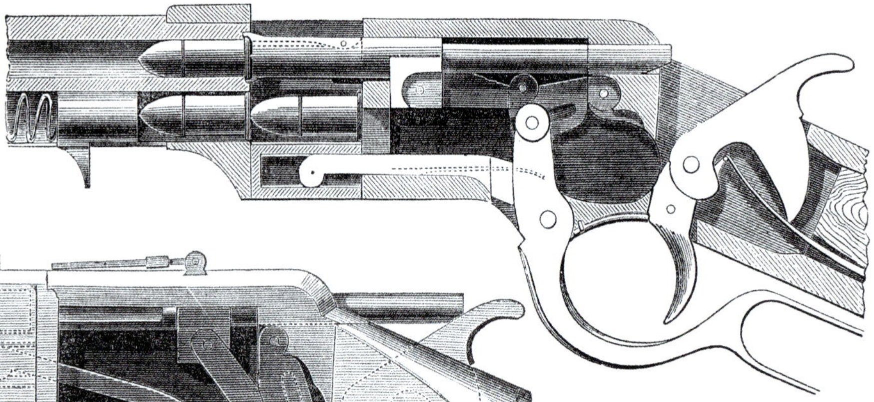

Sectional view of working parts with guard restored to place and ready to fire.
From Cleveland, *Hints to Riflemen.*

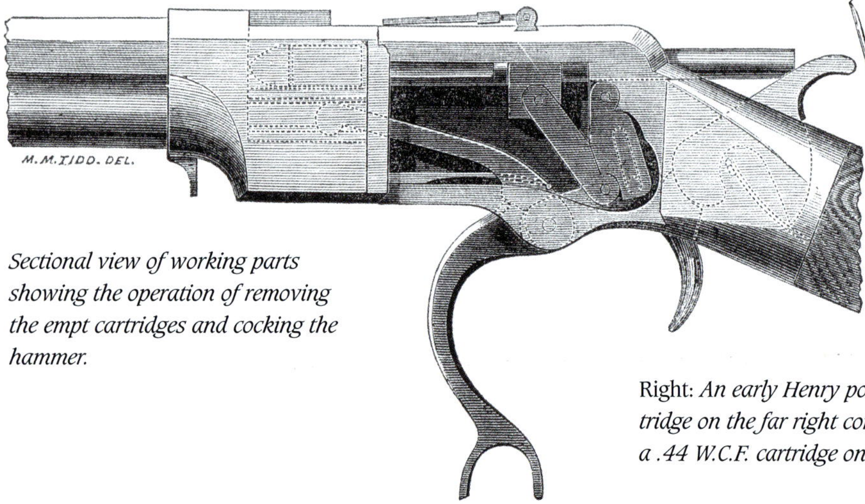

Sectional view of working parts showing the operation of removing the empt cartridges and cocking the hammer.

Right: *An early Henry pointed cartridge on the far right compared with a .44 W.C.F. cartridge on the left.*

known as the Model 1866 Winchester. Production began at the New Haven Arms Company in 1861 with a pointed bullet, which weighed 216 grains and had a 26-grain powder charge. Later, velocity was improved by retaining the same amount of powder but reducing the bullet weight to 200 grains. The Henry flat cartridge was later introduced in 1862. Their performance far exceeded that of the Smith & Wesson Volcanic ammunition.

By April of 1861, Winchester's personal finances had improved sufficiently to allow him to advance the company the money needed to begin construction of the tools and fixtures to begin manufacture of the Henry rifle. The company was in no position, however, to pay royalties to B. Tyler Henry for his improved rifle or to pay him for his cartridge design.[5] The new rifle was to bear Henry's name but, according to the custom of the time, and at Oliver F. Winchester's direction, his patent was assigned to Winchester with no direct monetary compensation. Henry agreed to give up a salary of $1,500 annually in exchange for a contract to produce 5,000 rifles using the New Haven factory, tools, ma-

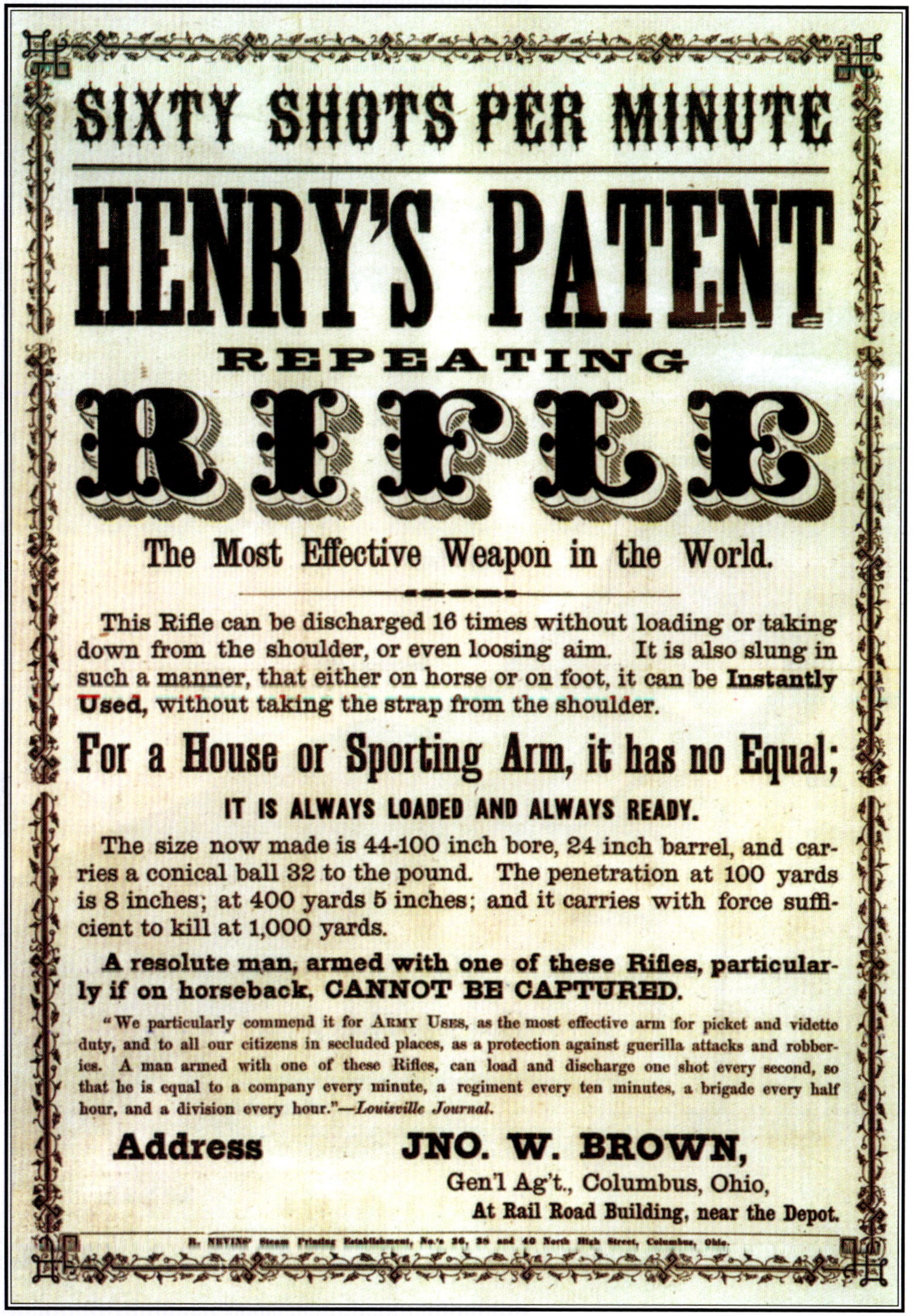

Carte-de-vista of a young Civil War soldier with his Henry rifle. Donovan Williamson collection

Carte-de-vista of young Civil War veteran who was probably mustered out with his Henry rifle.
Donovan Williamson collection

chinery and materials. Under this contract, all employees and their compensation were Henry's responsibility. When completed, the rifles were sold to the New Haven Arms Company at a fixed per-piece price, and the difference between Henry's costs and the contract price was his profit. It would prove to be a wise decision, for over the next five years, while the arrangements remained in force, Henry received $15,000 rather than the $7,500 that he would have earned on salary.[6]

The Henry rifle enjoyed great success during the Civil War. The U.S. Ordnance Department purchased approximately 1,700 Henry rifles in 1863 through 1865 for issue to the 1st Maine Cavalry Regiment and the 1st D.C. Cavalry Regiment. Over 200 of these Henry rifles were captured in September of 1864 from the 1st D.C. Cavalry Regiment and subsequently issued to the 7th, 11th and 35th Virginia (Confederate) Cavalry.[7]

Many authorities believe that most of the rifles manufactured before April of 1865 were purchased by Federal soldiers who appreciated their firepower. The Henry rifle was especially popular with soldiers from Illinois, Indiana and Kentucky regiments in the Federal Army of the Cumberland. Some western regiments like the 66th Illinois Infantry were primarily armed with privately purchased Henry rifles.

Oliver F. Winchester, who directed the firm's sales and marketing, had hoped to aquire some additional military contracts, but this never transpired. Instead, the Henry and its descendants, the Winchester rifles, would find fabulous success in the hands of pioneers, ranchers and hunters on the new post-Civil War American frontier. By the end of production, approximately 14,000 Henry rifles were produced.[8] Henry worked for the New Haven Arms Company until 1864 when the company was reorganized and moved to Bridgeport, Connecticut.

In May of 1865, with Winchester in Europe, Henry and the secretary of the New Haven Arms Company changed the company's name to the Henry Repeating Rifle Company by using a power-of-attorney Winchester had left with them. To avoid undue publicity, Winchester simply forced the bankruptcy of the new company by calling due the notes he held and denying them the manufacturing facilities at the Bridgeport factory. He then formed the Winchester Arms Company, soon to be renamed the Winchester Repeating Arms Company.[9] Henry decided to remain in New Haven where he operated a small gun store and machine shop until his death on June 8, 1898. His later years were spent in obscurity, and he was virtually unknown at the time of his death.[10] He left behind a great legacy in his rifle and its cartridge and turned the New Haven Arms Company into a successful venture, which ultimately led to the formation of the greatest firearms company of its time, the Winchester Repeating Arms Company.

THE IRON-FRAME HENRY

Some of the first Henry rifles manufactured had iron frames and buttplates and are referred to by collectors as "iron frames." The sight blade, carrier and cartridge follower were the only brass components used. The frame, buttplate, barrel and magazine were bright blued, and the hammer was case-hardened. There is no lever latch and no lever spur on these rifles. It is estimated that less than 275 guns were produced with an iron frame. Less than 100 are known.[11]

The First Model Iron-Frame Henry is highly prized by collectors. Why the New Haven Arms Company departed from the previous brass frame of the Volcanic rifles is unknown.

An early iron-frame Henry rifle, serial number 110.
Courtesy of Rock Island Auction Company

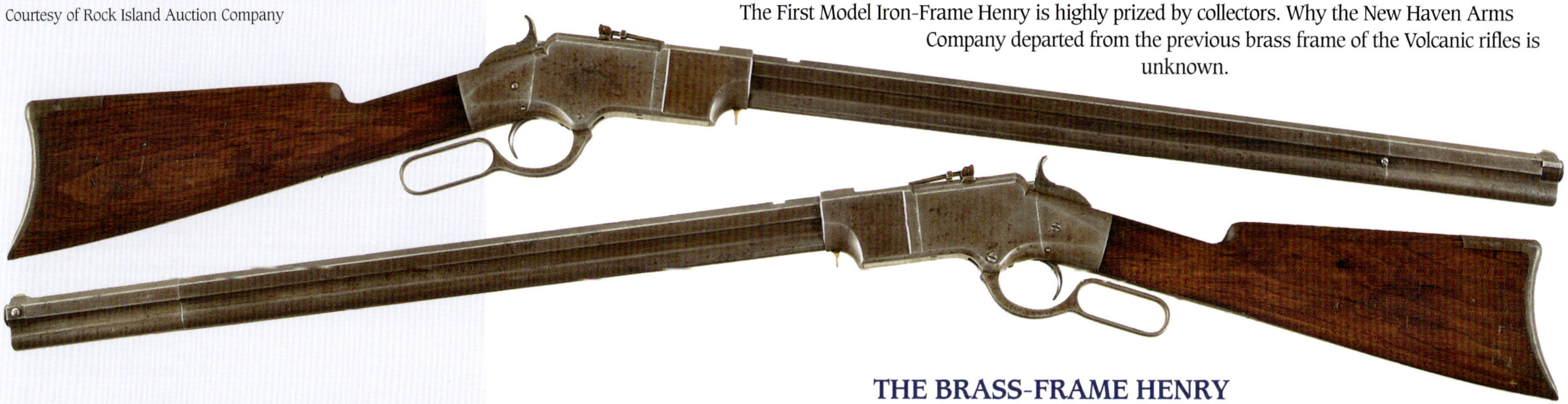

THE BRASS-FRAME HENRY

This First Model brass-frame Henry rifle was the most advanced rifle manufactured during the Civil War. It had the early Henry rifle features; a buttplate rounded at the heel, a 1,000-yard rear ladder sight mounted on the receiver, and it lacked a dovetail on the barrel for a rear sight. There was no sling swivel on the stock or sling loop on the left side of the barrel.

Early brass-frame Henry rifle, serial number 256.
Courtesy of the Rock Island Auction Company

FINALLY... THE WINCHESTER

The ongoing cartridge problem had been solved by Henry. The rifles, however, continued to have serious problems with regard to the cartridge magazine and magazine block. The open slot on the underside of the magazine allowed the entry of dirt and other foreign material and could be dented or damaged rendering the gun inoperable. In addition, the magazine block could become rusted fast to the barrel, preventing rotation for loading. If the T-bar catch became inoperative, the magazine block could rotate freely, cutting the magazine spring or causing its loss. Finally, the barrel with its magazine was simply too heavy, resulting in a rifle with poor balance. All of these problems were to be corrected by Nelson H. King who succeeded B. Tyler Henry as plant superintendent when Henry left the New Haven Arms Company. King, from Bridgeport, Connecticut, was listed in that city's directory as a machinist.[12]

Nelson H. King, circa 1885. Olin Corporation photograph

First Model 1866 carbine, serial number 13784, with an early receiver with Henry characteristics. There is no flare of the receiver at its forward end and hence, is called a "flat side" by collectors. There is also a flat loading gate. Manufactured in the first year of production, 1866. Note the early, non-adjustable rear sight. This sight is also found on early First Model rifles. It is thought that the First Model 1866 flat-side carbines were produced first, since the known examples all have serial numbers lower than the First Model rifles. Very few of the flat side rifles and carbines were ever manufactured. Randy Hughes collection

King's Patent No. 55,012 issued on May 22, 1866, whereby the magazine could be loaded through a port in the right side of the receiver, was a major improvement in firearms design.[13] The shooter no longer had to turn the rifle around and fumble with a magazine follower and magazine block. The rifle could now be loaded by simply inserting cartridges into the magazine through the loading port. This could be done in tight situations and even while on horseback. It also had advantages in that it could either be readily charged for a single shot or the magazine could be replenished at any time. The magazine was now a closed tube with a walnut forearm, which provided additional protection to the magazine. Initially, the forearm had no cap, though later, a brass fore-end cap was added to the rifles; the carbine forearm had no end-cap. With the reduced weight and more comfortable forearm, balance was improved and it became a more pleasant gun to shoot. Winchester paid homage to the two men who made the new rifle a success with a barrel marking that was used to approximately serial number 23,000:

HENRY'S PATENT – OCT. 16. 1860
KING'S PATENT – MARCH 29. 1866

King's recognition continued even after the inscription had been changed. From serial number 23000 to the end of production it read:

WINCHESTER'S REPEATING ARMS. NEW HAVEN, CT.
KING'S-IMPROVEMENT-PATENTED-MARCH 29, 1866. OCTOBER 16, 1860.

First Model 1866 rifle, serial number 15902, also showing an early receiver but with a grooved loading gate, uncapped, all-wood fore-end and saddle ring.
Randy Hughes collection

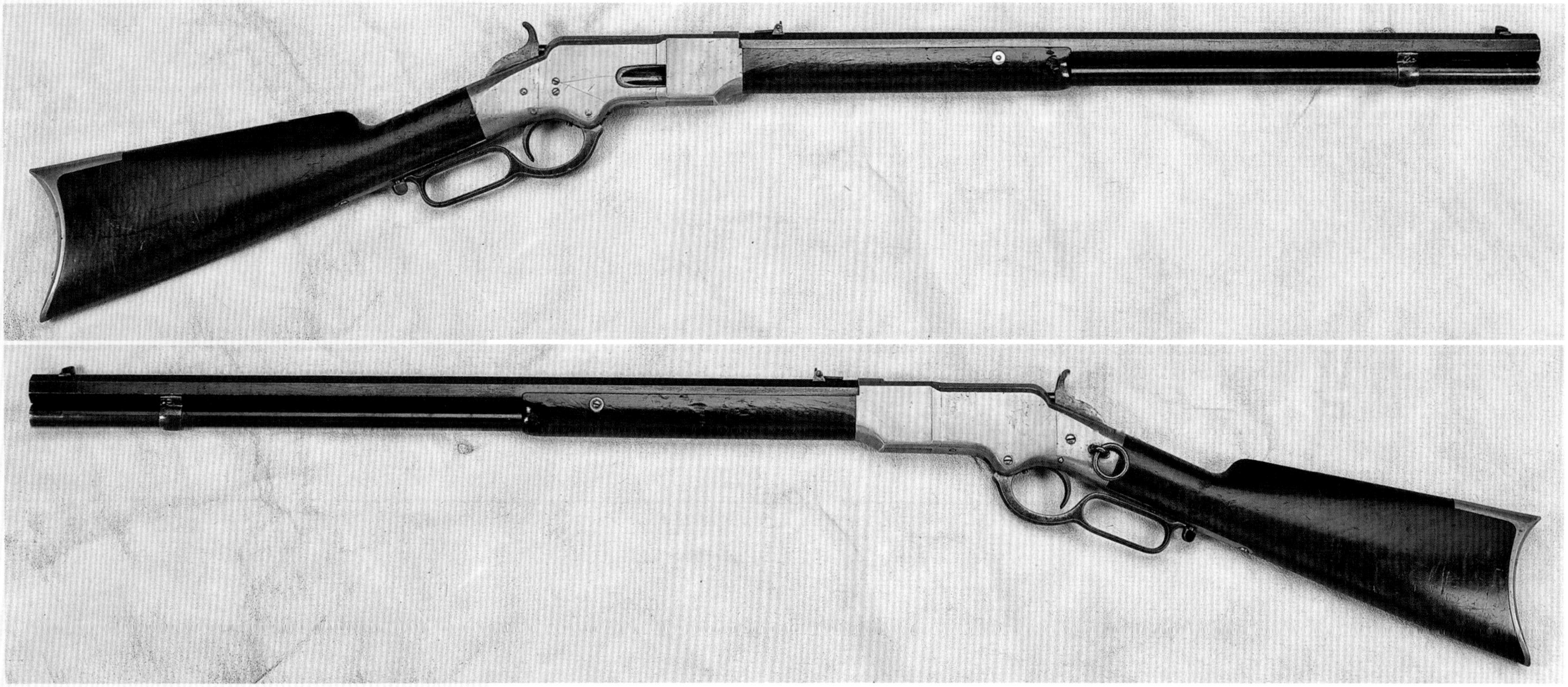

The final step in the pathway to the Winchester rifle had been taken. The "Improved Henry Rifle," as it was first called by the company, became the Model 1866 Winchester. The following year, on February 20, 1867, a new corporate name was given to the company — the Winchester Repeating Arms Company.

First Model 1866 rifles, in caliber .44 Henry RF, have a brass frame with no flare at the forward part of the receiver, have a brass buttplate and usually have a grooved loading gate, while First Model carbines have a flat loading gate. The early, all wood, uncapped and chamfered fore-end is attached by a lateral screw in a brass escutcheon. The left side of the receiver has a saddle ring as found on the early rifles. First Model barrels have less taper than the later models. The magazine retaining band is attached by screws rather than being dovetailed. Less than ten of the First Model 1866 rifles are believed to have been produced.[14]

At last, the path had reached its destination, the production of the first Winchester, the Model 1866. Further developments would lead to the Model 1873 and to the last of the toggle-link actions, the Winchester "Centennial" Model 1876. The lever action repeater had finally come of age, and its pathway had been paved by the Volcanic firearms.

A Second Model 1866 with a brass fore-end cap, serial number 21398. This style soon replaced the fore-end of the First Model.
Randy Hughes collection

epilogue

This story of the Volcanic firearms sprung from our interest in the Winchester rifle, beginning with John Parson's seminal book, *The First Winchester, The Story of the 1866 Repeating Rifle* and later with the introduction to the Volcanic firearms by George Madis in *The Winchester Book*. R. Bruce McDowell's *Evolution of the Winchester*, encouraged us to learn more of the history about the predecessors and the problems, and their solution, that led to the development of the Winchester rifle.

To these men we owe a great debt of gratitude for lighting a path for us to follow. We believe it is necessary and important to set down a record of these firearms, not only to illustrate their beauty but also to note the fact that they played a great role in future firearms development and, in their day, were used in many ways, and by many men not previously recognized nor acknowledged. Our travels, beginning with Walter Hunt and his Volition Repeater, and ending with the introduction of the Model 1866 Winchester rifle, has accorded us great pleasure with the discovery of many new facts and the opportunity to meet many new friends. However, as Shakespeare said in The Tempest, "...what's past is prologue; what to come, is yours and my discharge." It is in this spirit that we welcome the additional knowledge that other students of these firearms will uncover in future years.

Edmund Lewis and Stephen Rutter, 2010

appendix I

AMMUNITION AND FIREARMS PATENTS

Much of this material has been researched, investigated and explained before, most notably by R. Bruce McDowell, and it has been the mission of the authors to collect, collate and further explain their concepts as well as add new findings and explanations of several important areas in the development and manufacture of these firearms. This array of patent drawings and letters includes the most significant patents that were issued along the pathway from Hunt's Volition Repeater to the development of the improved Volcanic — the Henry Rifle. They outline the thoughts and dreams of men attempting to perfect the concept of a breechloading repeating firearm and its ammunition. All of these inventors played a part:

Smith & Wesson.
Cartridge.
No. 11,496. Patented Aug. 8, 1854.

SMITH & WESSON.
Cartridge.
No. 14,147. Patented Jan. 22, 1856.

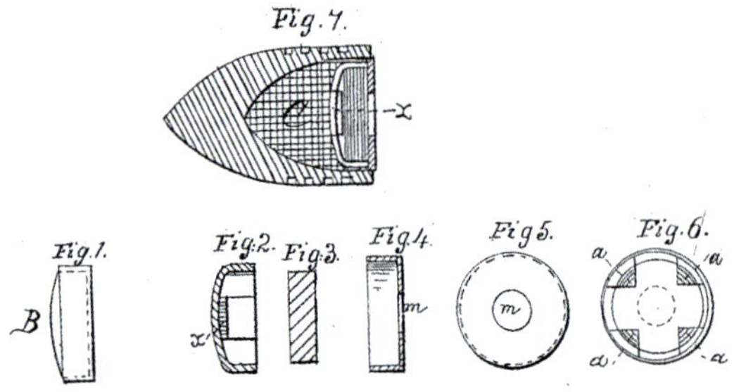

W. HUNT.
Cartridge.
No. 5,699. Patented Aug. 10, 1848.

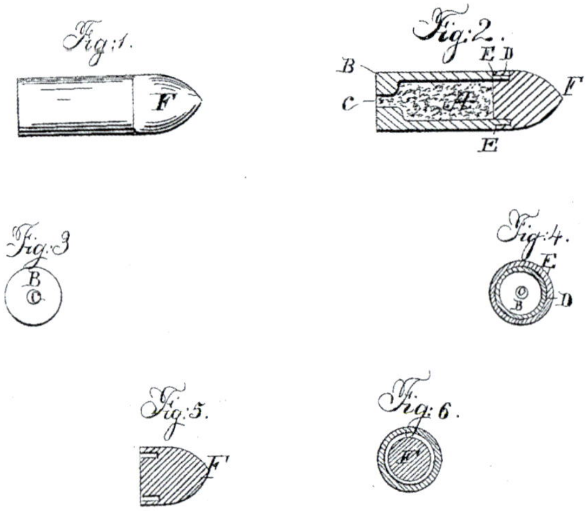

W. HUNT.
Cartridge.
No. 5,701. Patented Aug. 10, 1848.

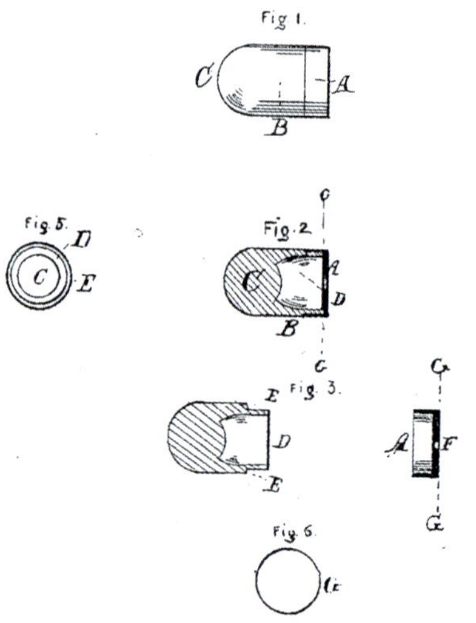

No. 6,663. PATENTED AUG. 21, 1849.

W. HUNT.

COMBINED PISTON BREECH AND FIRING COCK REPEATING GUN.

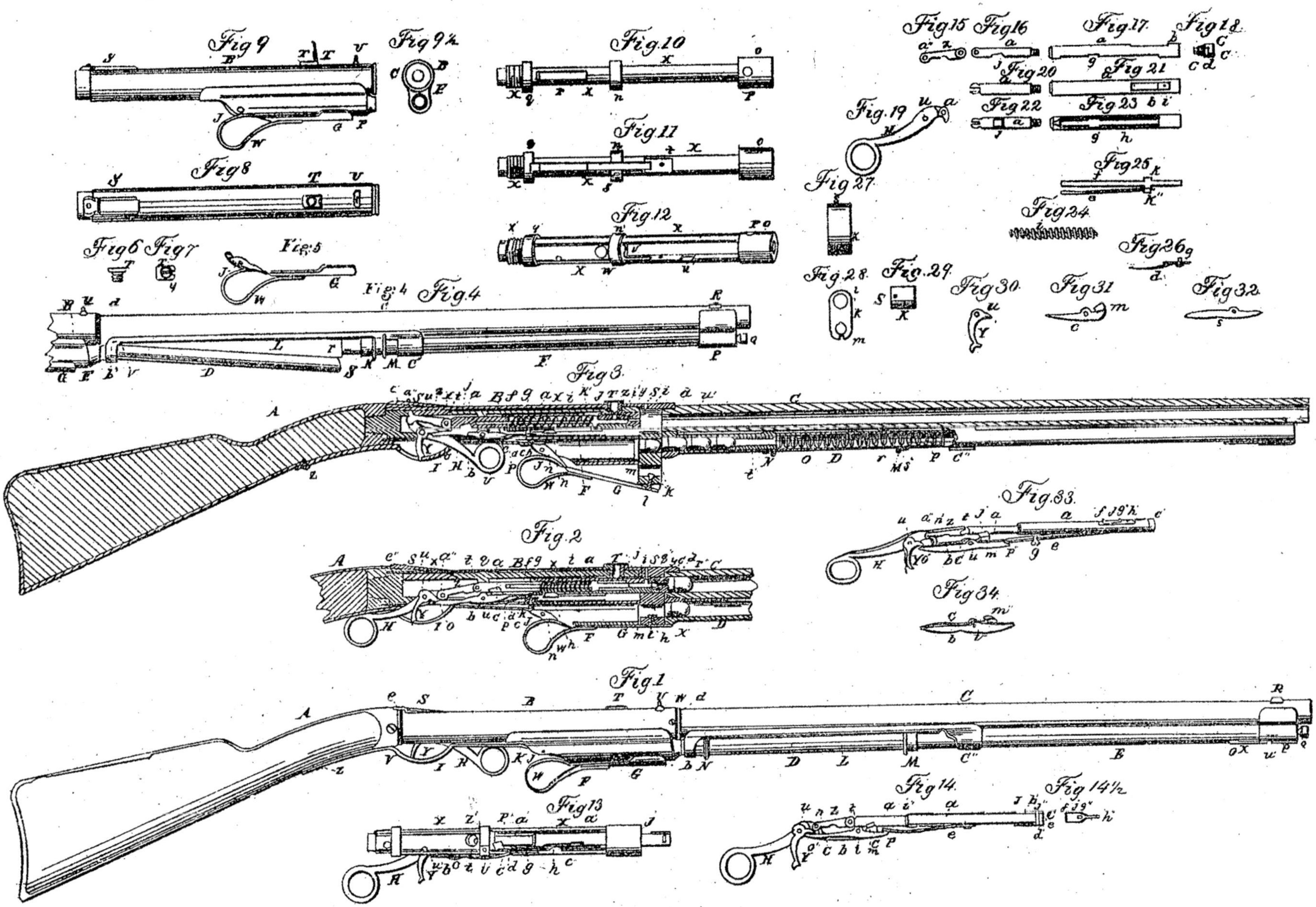

L. JENNINGS.

BREECH LOADING FIREARM.

No. 6,973. Patented Dec. 25, 1849.

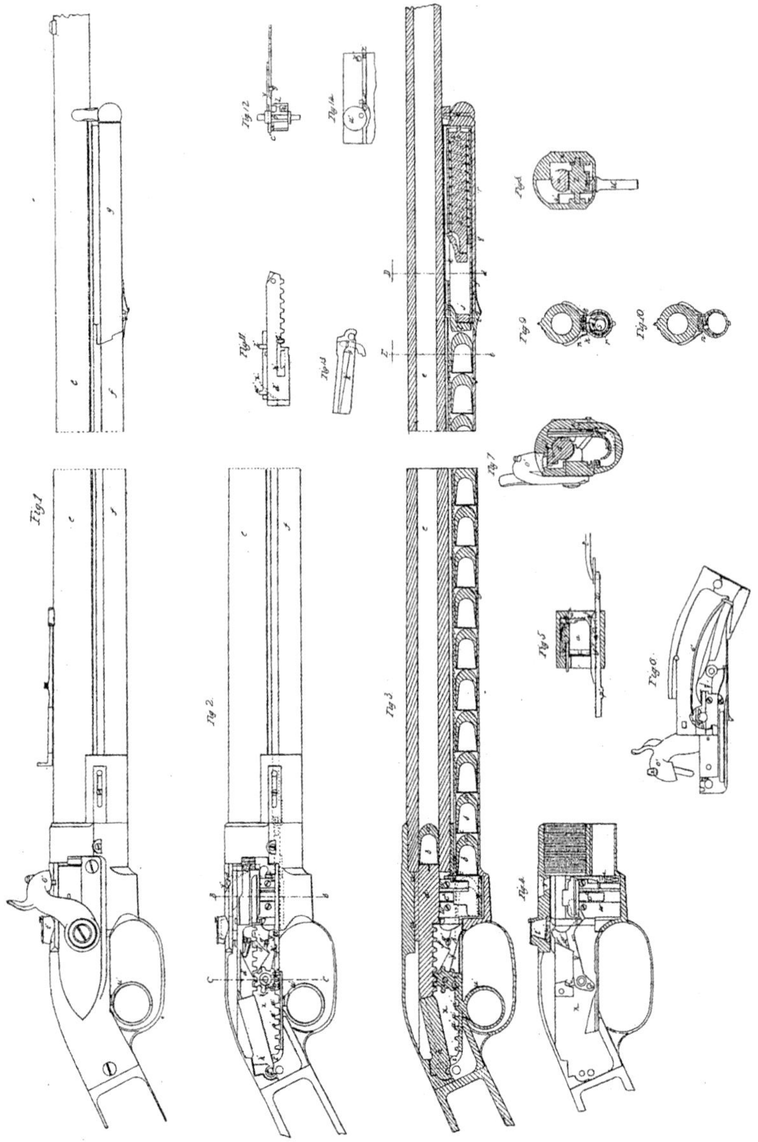

H. SMITH.

Magazine Fire-Arm.

No. 8,317. Patented Aug. 26, 1851.

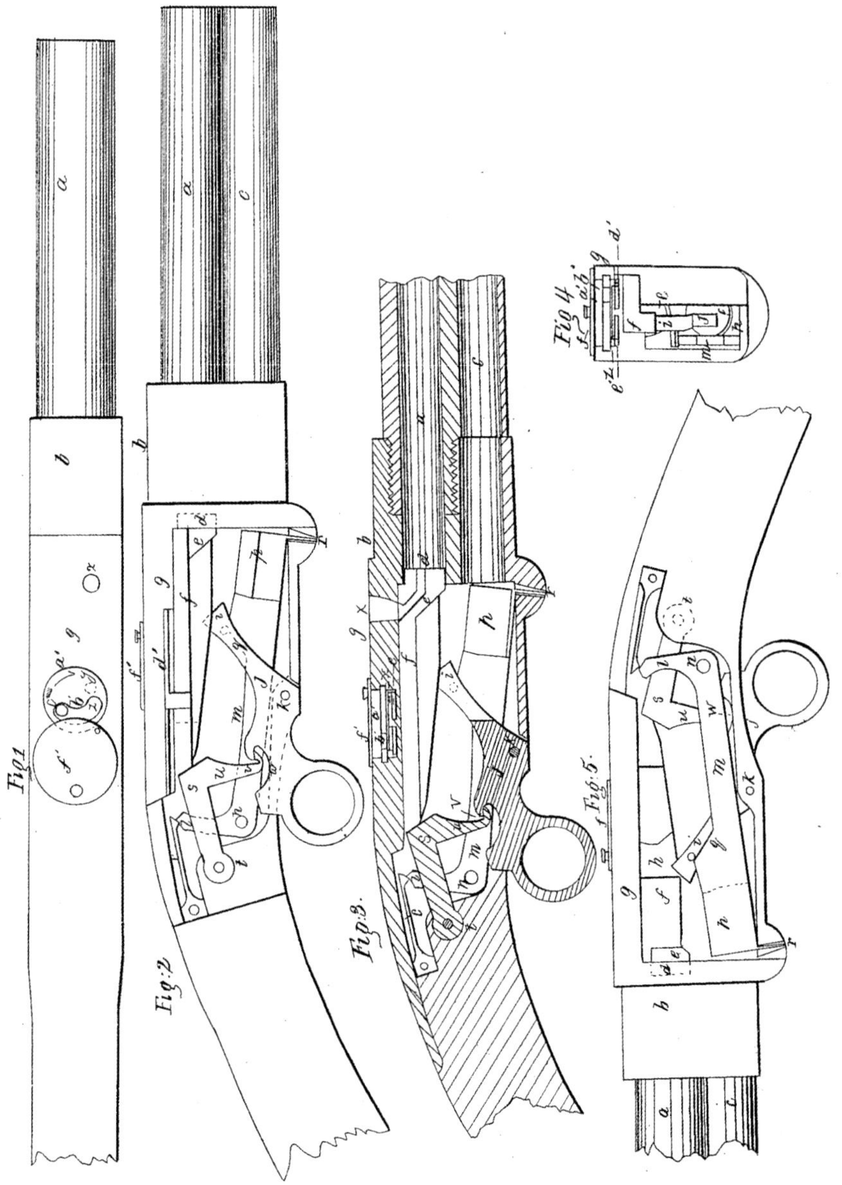

2 Sheets—Sheet 1.

H. SMITH & D. B. WESSON.

MAGAZINE FIREARM.

No. 10,535. Patented Feb. 14, 1854.

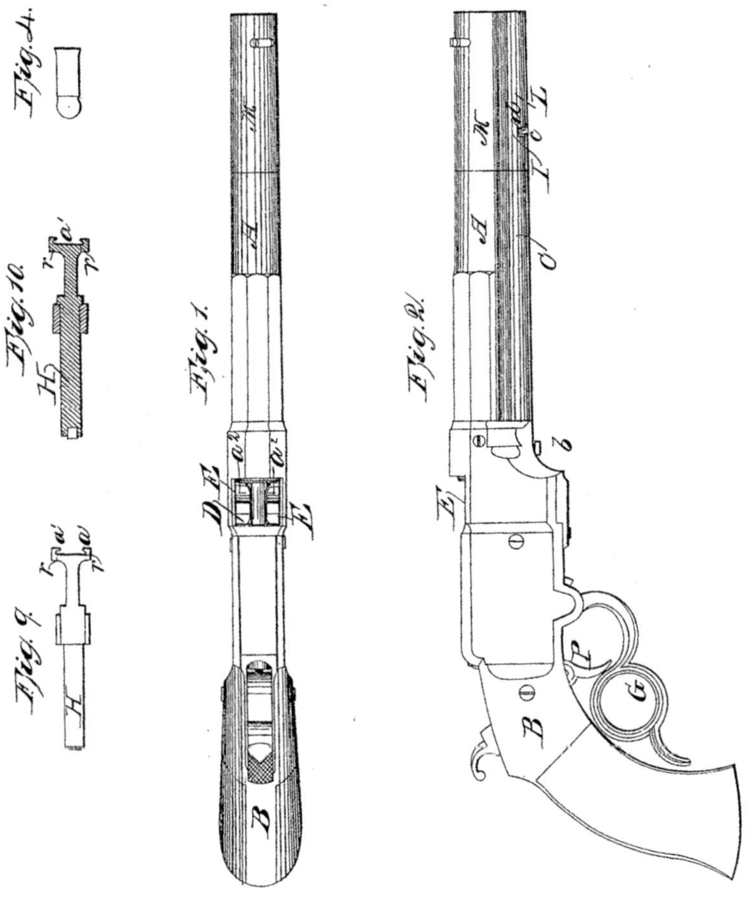

2 Sheets—Sheet 2.

H. SMITH & D. B. WESSON

MAGAZINE FIREARM.

No. 10,535. Patented Feb. 14, 1854.

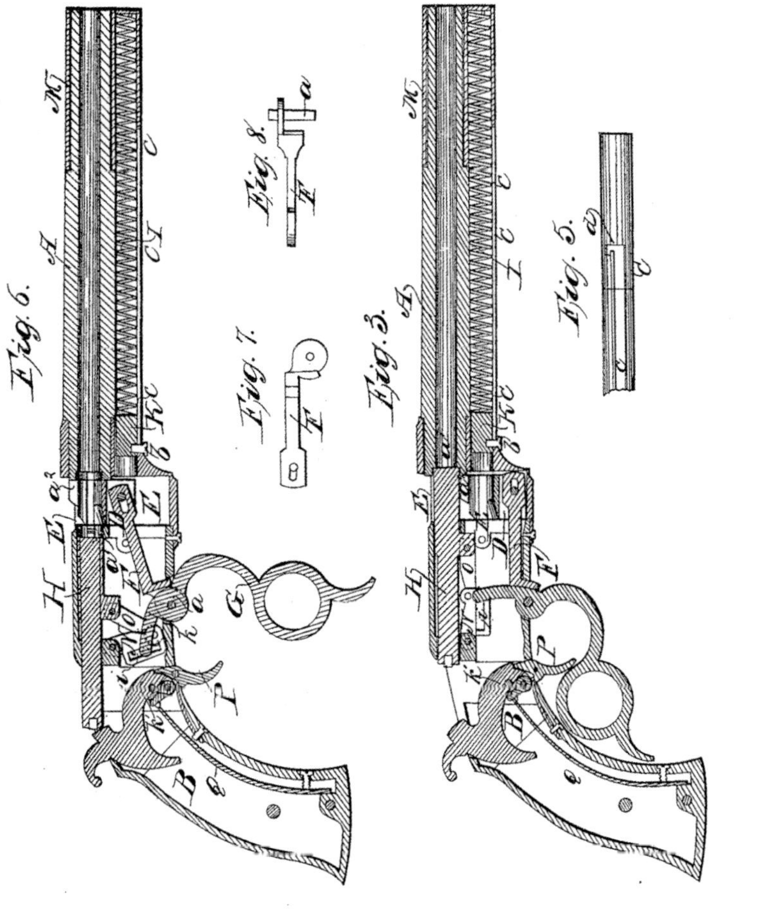

W. C. HICKS.

NIPPLE FOR DISCHARGING OR WITHDRAWING CARTRIDGES FROM BREECH LOADING FIREARMS.

No. 16,797. Patented Mar. 10, 1857.

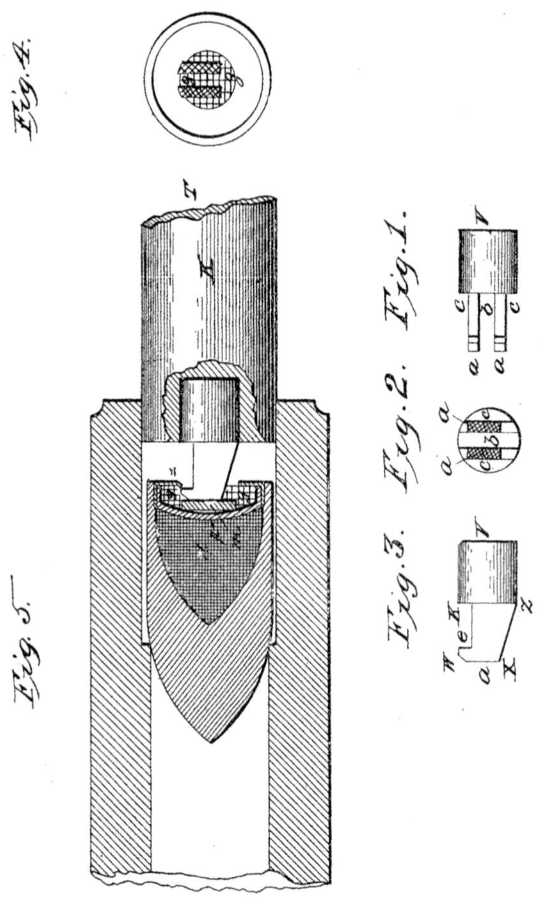

No. 27,933. PATENTED APR. 17, 1860.

H. SMITH & D. B. WESSON.

METALLIC CARTRIDGE.

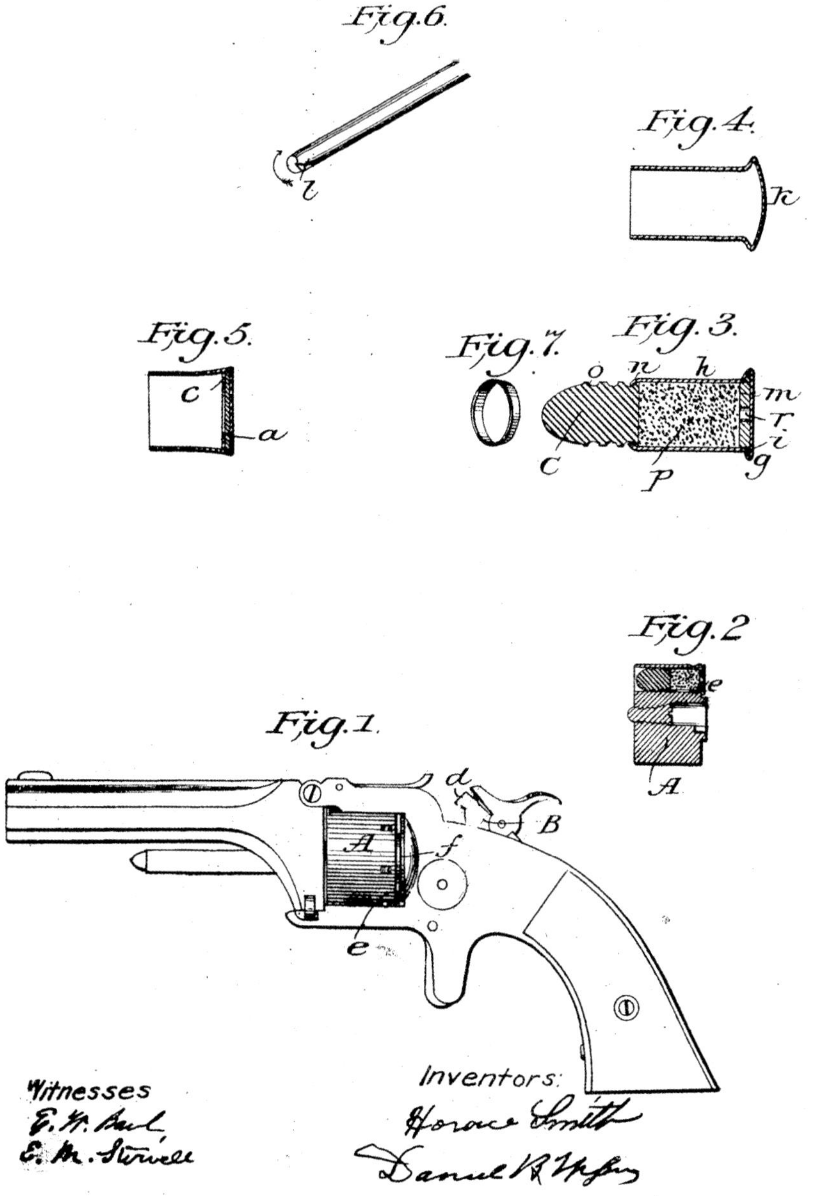

B. T. HENRY.

Magazine Fire Arm.

No. 30,446. Patented Oct. 16, 1860.

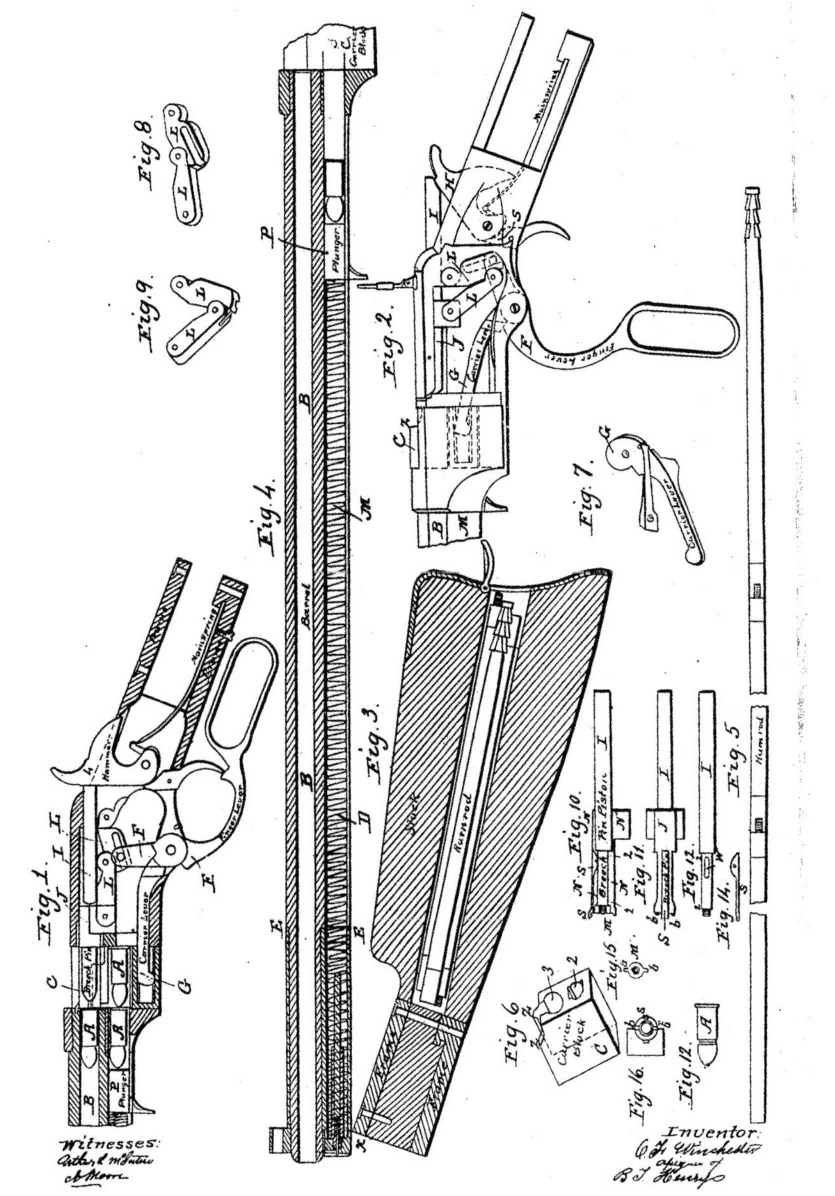

N. KING.

Magazine Fire-Arm.

No. 55,012. Patented May 22, 1866.

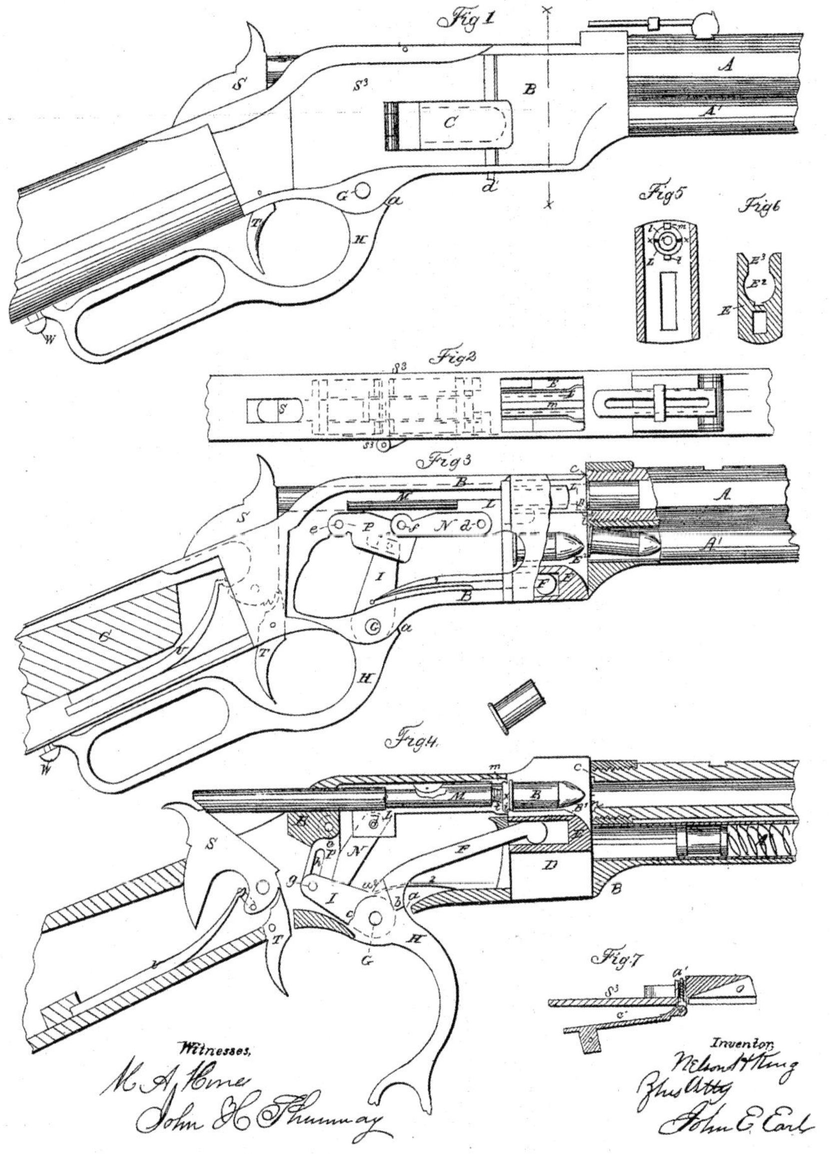

appendix II

OLIVER WINCHESTER'S TESTIMONY REGARDING THE VOLCANIC ARMS AND THEIR AMMUNITION

There is a wealth of historical material related to the Volcanic firearms that can be found in the archives of the McCracken Library at the Buffalo Bill Historical Center in Cody, Wyoming. Thomas E. Hall, past curator of the Winchester Gun Museum when it was located in New Haven, Connecticut, published some of this information in the *Gun Digest* in 1958. Interesting information concerning Smith & Wesson and the Volcanic Repeating Arms Company can be found in the bound copy of *Morse Arms Manufacturing Company versus Winchester Repeating Arms Company/Complainants Main Case/'Extracts from the testimony of Oliver F. Winchester in Renwick against Pond, offered in evidence by Complainants Counsel, December 8, 1871*.

The comments made by Oliver F. Winchester during the court proceedings relate facts regarding the manufacture of Volcanic pistols, including their production figures and a highly detailed description of the Volcanic cartridges, including variations. In these proceedings, we learn that Oliver Winchester not only had an intimate and accurate working knowledge of the construction and production of these firearms, but that he also had personally fired them and was, himself, well aware of their deficiencies. Winchester relates in his own words that:

"...After the organization of the Volcanic Repeating Arms Company they purchased of Smith & Wesson, in the month of August, 1855, all their machinery, tools and fixtures, and arms, finished and unfinished and removed them from Norwich to New Haven the same month. Among these assets purchased were about 300 pistols, of both sizes, finished and unfinished, and the parts of some 280 to 300 more pistols, not assembled. The company was occupied, in part, nearly or quite a year in overhauling and repairing and assembling these 600 pistols, more or less, and manufacturing a new model and the tools and fixtures for that model. The company (Volcanic Repeating Arms Company) continued for the remaining six months of its existence in finishing and selling pistols of this new model until the 3rd of February, 1856, when they passed a vote to go into insolvency and made an assignment accordingly".

This is the only direct information that historians, most notably Thomas E. Hall, have been able to obtain concerning production figures of the "Volcanic" type arms made by Smith & Wesson. Apparently, at least some 600 pistols of the styles originally made in Norwich were produced (Tom Hall, Part II, *Gun Digest*, 12th edition, 1958, pp. 122, 123). Winchester continued his testimony and when asked to describe the construction and mode of operation of the pistols, finished and in progress, which the company purchased from Smith & Wesson, related in great detail the construction and operation including that of the "V" nipple (the hooked extractor), which was completed and "perfected" in July of 1856. Winchester stated that: *"Mr. Hicks' sanguine representations led the directors to hope it would prove a cure for the evil. Subsequent trials proved it to be a failure, and it was abandoned."*

He then describes in great detail the construction and function of the newly patented Hicks extractor with a double hook and added: "The twin nipple, so called, proved equally defective..."

Further he stated that:

"Mr. Hicks, about that time, or soon after, presented to us another supposed improvement, consisting substantially of two prongs or projections on the end of the nipple, each having the forward end or point substantially like the single nipple previously in use by the company, as shown on Exhibit No. 6, except that, owing to the limited space in which they were to operate, they were each made much thinner at the base. This, after trial proved a failure, and was abandoned. The cause of the failure of the 'V nipple' was owing to the flatness of the forward end which, instead of cutting and penetrating the cork simply, in most cases compressed the cork over the fulminate, thus forming a cushion over the fulminate to protect it from the blow, and thus prevent an explosion; and for the same reason, i.e., pressing on so large a surface of the cork, would force the ball forward, or too far into the chamber, so as to be beyond the reach of the firing pin. The hooks of this nipple, when engaged with the raw edge of the metal disc, still had a stronger hold than a single nipple, but would fail to extract the ball when it missed fire, because of the friction between the surface of the ball and the surface of the chamber required a greater force to withdraw the ball than it did to withdraw the primer; and thus the primer was pulled out and the ball and powder left behind. The twin nipple, so called, proved equally defective from other causes:

First — *Being necessarily made very slender, they were easily bent, and thus, as I have seen, many of them became inoperative.*

Second — *The forward end being nearly sharp, or cutting edges, would often, after firing, retain the piece of cork, cut out between the two points; thus, when the piston was moved forward to fire the next charge, , this piece of cork between the two nipples would prevent the entering of the points into the primer, by acting as a cushion or stopper.*

Third Another objection, similar to that against the 'V nipple' was that two points upon the end of the nipple required a greater force to drive them through the cork and the hooks past the metal ring, than one required; consequently the ball was apt to be carried too far forward to insure the firing, say half a hundredth of an inch too far, which was often sufficient to prevent the firing pins from reaching the anvil on which the fulminate rested, and so failed to insure ignition. Therefore, to remedy this, by increasing the friction between the surface of the ball and the surface of the chamber, by enlarging the diameter of the ball, failed of success, as it increased another evil called thimbling , the forward end of the ball being blown out, the rear end or base of the ball remaining in the chamber.

There was no sufficiently positive means of stopping or holding the ball at a fixed point in the chamber; the shoulder in the chamber, which was, at one time, relied upon for that purpose, proving insufficient, for the reason that the diameter of the chamber corresponded with the diameter of the bore of the barrel, measuring from the bottom of the grooves, five in number; thus leaving to form that shoulder, simply the ends of the lands, which, being only 1/100th inch in depth, formed no certain stop for the ball, but cut into, with slight pressure, the soft lead of which the balls were formed, so as to make it entirely unreliable for that purpose...it was hoped it would serve.

Q. 4. At the time that the Volcanic Arms Company made the purchase from Smith & Wesson, as mentioned by you, did the company receive any balls from Smith & Wesson; and if yes, what kind or kinds?

A. They did; of two kinds. I will state that Daniel B. Wesson was one of the directors, appointed at the first meeting for the appointment of directors of the Volcanic Repeating Arms Company. With the purchase from Smith & Wesson came a small quantity of balls of two kinds; one kind had a primer in the base, like Exhibit No. 7, formed by a metal disc resting upon the powder, the fulminate pill resting on the centre of that disc, and held in its place by a thin circular disc of cork, with the diameter designed to fit tight into the base of the ball; the whole thing being held in place by pressing the lead of the ball snugly around, and slightly over the edge of the cork. Of this kind we had several hundreds, perhaps thousands; I know they were lying about for years. The other kind, as shown in Exhibit No. 5, were substantially of the same construction, except that the parts of the primer, the disc, fulminate, and cork, were enclosed in a brass cup having a small circular hole through the closed end, which was pressed down and doubled over on the inner edge, so as to bind the whole together in one piece. Mr. Wesson as a director of the company, called the attention of the board especially to this last named ball, Exhibit No. 5, explain the object of its construction, the difficulties it overcame, and advised immediate arrangements for procuring a supply. The meetings of the company were weekly, generally; I think at each of the first three meetings held by the directors after organization this was a subject of consultation. Such arrangements were immediately made, in consequence of Mr. Wesson's recommendation, with Crittenden & Tibbals of South Coventry for a supply. It was the cartridge thus adapted by the company, and no other was used by them to my knowledge, during its existence.

Q. 5. Have you any of the pistols which the company purchased from Smith & Wesson at the time of the formation of the company; and if yea, will you produce them?

A. I have probably twenty or more, of which I now produce one, to be marked by the Examiner Defendants Exhibit No. 9

Q. 6. About how many pistols like Exhibit 9 did the Volcanic Repeating Arms Company sell prior the time when Mr. Hicks commenced his experiments with a view to improve the arm.

A. I can speak with no certainty on that point now; I have a monthly account of sales by the agents during those months; but not anticipating this question I am not prepared to answer.

Q. 7. Give the number to the best of your recollection?

A. Not less than one hundred, probably over three hundred.

Q. 8. During that time did the company sell, and if so, about how many, pistols of a different size from Exhibit No. 9?

A. My understanding of the previous question was that it referred to all sizes of pistols like the Exhibit No. 9, and my answer was not confined to the large size.

Q. 9. Look at Defendant's Exhibit No. 4, and state whether it has ever been in your possession, and from whom you received it, and when.

A. I recognize this pistol as one that I received by express, direct, from Mr. John W. Boynton, of Hartford, accompanied by a letter; the date I cannot fix, because I have not that letter with me, but it was since the commencement of this suit. On this receipt, I immediately sought for the manufacturer's regis-

tered mark or number, so as to be able to identify it, and approximately to determine the times of its manufacture; the mark is "C", which, by the Smith & Wesson system, would determine it to be the 311th pistol of that construction made by them.

Q. 10. Has the nipple been changed in Exhibit No. 4, since you received it?

A. It has not been changed or altered and the pistol is in every respect, to the best of my knowledge and belief, in the same condition as when it came into my hands.

Q. 11. Were any of the arms with the V-shaped, or with the double nipples, sold by the company; and if so, how many?

A. The company sold nothing directly. All arms were sold by their agents in New York, Wheeler & Post, and I have no means of answering the question positively, as their reports are not here, and probably would not distinguish, if they were here, what were single and what were double.

Q. 12. Did the company make any single nipples after they had tried the V nipples and the double nipples?

A. They took out double nipples, and substituted single nipples They may have been the same single nipples that were taken out before for the substitution of the double.

Q. 13. Does the Winchester Repeating Arms Company, of which you are President, manufacture the arm like Plaintiff's Exhibit B, and which was sold by the defendant in this suit?

A. We manufacture that arm, and other forms of it, too.

Q. 14. Does the company manufacture such under any patent or patents? And if so, to whom, and when?

A. We manufacture that arm under the patent granted to B. Tyler Henry, October 16th, 1860, and also under a number of patents granted subsequently, which, however, have no direct bearing, in my judgment, upon the question in dispute.

(Defendant's counsel offers in evidence a certified copy of the Letters Patent referred to in the last answer of the witness, which is marked Defendant's Exhibit No. 10)

Q. 15. When was the manufacture of such arms commenced?

A. The question is rather general. The model or first gun was manufactured as the result of improvements during the latter part of the year 1858, and all of 1859, and the early part of 1860, when the model was completed and forwarded to Washington with an application for a patent, which was granted in October of that year. The commencement to manufacture quantities (of the Henry rifle) was made late in 1861."

Hall, Thomas E., "Forerunners of the First Winchester," *Gun Digest,* 12th Edition, Part 2 of 3, The Gun Digest Company, Chicago, Illinois, 1958.

appendix III

VENDITTI CONTROVERSY: THE PISTOLS OF HORACE SMITH, DANIEL B. WESSON and PIETRO VENDITTI

Some maintain that the Venditti repeating pistols were manufactured in Naples, Italy, between 1875 and 1880.[1] Others believe they were produced near Lancusi, Italy, a village on the outskirts of Salerno, possibly as early as 1843.[2] These differing opinions result in the conundrum which persists to this day: did Venditti copy the Smith & Wesson design or was the Venditti design the inspiration for Smith & Wesson's lever repeating pistol?

Pietro Venditti was born on February 9, 1828, in the township of Cerreto, a province of Naples. Though the date is not noted, it is known that Venditti was condemned to an indefinite period of labor for killing a co-worker. He was in prison until 1860 when he was reprieved. He then worked at the state arsenal at Lancusi where he advanced quickly to become foreman. It seems that this was Venditti's first exposure to firearms. In 1871, at age 43, Venditti displayed a cannon model and

Barrel address of the First Model Venditti. Roger Quisenberry collection

First Model Venditti Pistol with two auxiliary magazine tubes in addition to the true under barrel magazine. Courtesy of Czerny's International Auction House, Sarzana, Italy

Second Model Venditti pistol and variation of the barrel address. Gerald Klaz collection

small breechloading pistols at the International Maritime Exhibition at Naples. On February 8, 1872, he registered his first Kingdom of Italy patent covering a small double-barreled breechloading pistol that fired Rocket Ball cartridges, and in 1873, Venditti took part in the International Exhibition in Vienna, displaying two cannons. It wasn't until February 27, 1875, that he registered a three-year Kingdom of Italy Patent, No. 77, for the first model Venditti pistol, a 26-shot repeater also designed to use a Rocket Ball cartridge similar to the Smith & Wesson No. 2 Volcanic cartridge. This pistol was similar to a Volcanic pistol, but it had three magazine tubes; one under the barrel and two additional ammunition chambers parallel to the barrel, on the right and left sides, designed to load the actual feeding magazine. This same idea was tried and discarded on the Henry rifle between 1862 and 1863.[3]

Venditti secured a three-year Kingdom of Italy Patent, No. 105, on March 15, 1877, for his second type Venditti pistol. It fired a cased cartridge rather than the Rocket Ball and had only one underbarrel magazine tube. Several transition models have been noted that have the barrel address of the first model with a single underbarrel magazine and the receiver frame of the second model.

A mention of the Italian politics and government of this period is in order, since the patent process changed. Prior to 1861, when the Kingdom of Italy was formed, only Bourbon patents had been issued, and all records of these were destroyed in 1860 with the overthrow of the Royal House of Bourbon.[4] Some of these pre-Unification patents were subsequently transferred. However, if Venditti's Patent Numbers 77 and 105 were actually transfers of prior Bourbon patents, it is strange that the Venditti repeating pistols were not exhibited at either the 1871 or the 1873 Exhibitions. Likewise, if these Kingdom of Italy patents were such transfers, why did he wait until 1875 and 1877 to submit them, leaving his ideas unprotected by patent during the five and seven years that the new Kingdom of Italy patents were available to him?

The controversy seems to have begun with a highly speculative article by Graham Burnside in the December 1958 issue of *The Gun Report*, in which he concluded: "Logic and common sense tend to say that the foreign gallery pistols that look like the Volcanic are not copies but rather the actual source of the Volcanic itself!"[5]

In 1962, William Edwards perpetuated Burnside's claims that Horace Smith attended the London Great Exposition of 1851.[6] However, Frank Sellers, in his 1973 article, refuted this statement saying that records exist to show that neither Smith nor Daniel Wesson left the country to attend the 1851 London Exposition.[7] Venditti was not likely at the Great Exposition either, because neither he nor a revolutionary device such as the Venditti repeating pistol were mentioned in the 320-page *Catalogue of the Great Exhibition*, printed for the Royal Commission in 1851. Venditti is likewise not listed in H.J. Blanch's *Century of Guns* that contained a list of makers who exhibited there.

Marcello Terenzi, in a paper presented to the American Society of Arms Collectors on April 2, 1970,[8] differed in conclusions from a similar work published in December of 1970 in the *Deutsch Waffen-Journal* by Emanuele Marciano and Gugliemo Torrello titled "Die Vorganger der Winchester."[9] In these articles, the authors conjectured that Venditti obtained a Bourbon patent, probably in 1848. They further summarized that Venditti displayed his pistol at the 1851 Great Exhibition in London and that Horace Smith (or Daniel B. Wesson) attended the Great Exhibition and purchased (or stole) the rights to the Venditti. Both Terenzi and Marciano-Torrello agree that the Venditti pistols were first made by Venditti in association with two other gunsmiths, Gaetano Papa and later Raffaele Papa, listed in a genealogical table of gunsmiths of Lancusi, a suburb of Salerno. This is supported by Gaetano Papa's initials commonly being found under the grips of some the pistols themselves.[10] Terenzi's published genealogical chart lists Gaetano Papa as having lived from 1846 to 1895 and Raffaele Papa from 1855 to 1924.[11] The dates that Venditti's collaborators lived refute the 1840s time period for production of the known Venditti pistols, and Venditti himself would have been only fifteen years old in 1843.

In response to Burnside's and Terenzi's conjecture concerning the origins of the Venditti pistol, and to correct what he felt was misinformation, Emanuele Marciano, the distinguished editor of *Diana Armi*, one of Europe's two finest monthly magazines dedicated to arms studies, wrote an article titled: "The Final Word on the Venditti-Volcanic Controversy," published in the 1973 *Arms and Armor Annual*.[12] In this thorough treatment of the topic, Marciano traces the documented facts of the early life of Pietro Venditti and concludes that Venditti's first interest in firearms followed his imprisonment, which lasted until 1860, fully six years after the appearance of the Smith & Wesson lever pistol.[13] Marciano mentions a number of other Italians who claimed to have invented design characteristics of the Volcanic and Venditti between 1877 and 1888.[14]

The Venditti pistol utilized the same toggle-link lever system depicted in Smith & Wesson's patent of 1854, and the external appearance, especially of the second model Venditti, mimics the pistols produced, not by Smith & Wesson Company, but by the later New Haven Arms Company. The barrels were blued, integral with the true magazine and, like the New Haven Arms pistols, had no side ribs.[15] The second model .40 caliber Venditti was built on a receiver the size of the small-frame New Haven Arms Company pistol.[16] The rotation of the magazine block loading varied also, with the Smith & Wesson, Volcanic and New Haven sleeve rotating clockwise to open, when viewed from the muzzle, and the Venditti rotating counterclockwise. The Venditti muzzle collar is bulkier, and only one stop was provided to limit the rotation, such that with rotation of the sleeve in the wrong direction (clockwise), the follower and spring are allowed to shoot out of the sleeve spring chamber and become lost.[17] The first model Venditti was designed to fire a loaded ball similar to the Volcanic Rocket Ball cartridge. The second model Venditti was designed to fire a centerfire metallic cartridge having a flanged base and, therefore, it had cartridge extractors added to the breech bolt. The chamber and carrier were made longer to accommodate the cartridge, and there were no auxiliary magazines. Both models had brass receivers and steel barrel-magazine units, but unlike the Volcanic pistols, the sideplates were also made of steel. Venditti pistol barrels, frames, sideplates and backstraps were engraved with a scroll-and-leaf pattern. The two-piece hardwood grips were "bag shaped" and generally checkered with a swivel and lanyard ring at the butt for the first model. A few second model pistols have flared grips with steel butt caps, longer than the standard grips.[18] A few transition models have been identified, having the first model barrel markings, smaller carriers, long pointed firing pin and no cartridge extractors. However, they lack the two auxiliary cartridge tubes, and never had them as evidenced by the receiver being of the second model type.[19]

It is estimated that Venditti himself, during 1875 and 1876, produced fewer than 100 of the first model pistols.[20] The first model Venditti pistol had its barrel address on the top flat, "SISTEMA VENDITTI", with a small circular mark containing a crown, a sword, a star and the letters "T.A." It is believed that the latter were a proof stamp of the arsenal of Torre Annunciata, near Naples where, after the 1870 Unification, civilian arms were probably proved and stamped.[21]

The second model Venditti's barrel address differed from that of the first model. The "E C1" indicated that he had formed a company or partnership and had possibly acquired financiers.[22] A few second model Venditti pistols have no barrel markings for unknown reasons.[23] The estimated number of second model Venditti pistols produced is fewer than 250.[24] A number of slight variations have been observed among second models, presumably due to them having been manufactured in several different shops or, possibly, they represent patent invasions.[25] Further details of these pistols may be found in McDowell's *Evolution of the Winchester*.[26]

The Venditti pistols do not demonstrate a progression of improvements, as seen in the Smith & Wesson lever pistols. The large-frame Smith & Wesson pistols began with a flat muzzle, no collar and a flat lock spring. As production progressed, they went to a crowned muzzle, short muzzle collars and a T-bar lock.[27] The Smith & Wesson barrel was a combination of octagonal at the receiver end transitioning to round at the muzzle and rifled with five grooves. The magazine tube was a separately attached piece, and the two ribs were likewise separate pieces, likely added as a buttress to strengthen the magazine-barrel junction. This barrel configuration of the Smith & Wesson is seen to have evolved when the Volcanic Repeating Arms

Company produced full-length octagonal machined barrels of a single piece of steel with the rib detail simply machined in during their short production period of 1855–56. This then progressed to the barrels of the New Haven Arms Company that were identical to their predecessor company except that there was no rib at all. The Venditti barrel was of one piece of steel with no rib and rifled with four grooves. It should be apparent, then, that the similar pistols are not the Venditti and the Smith & Wesson but the Venditti, and the New Haven Arms Company pistols. Therefore, to conclude that Smith & Wesson copied the Venditti toggle-link design, without copying the other barrel and magazine features seems highly improbable.

Both the Venditti pistols and the Volcanics were blind alleys in the history of firearms, and their exact origin should not rob these rather appealing, somewhat hapless weapons of their just historical and collecting merits.[28]

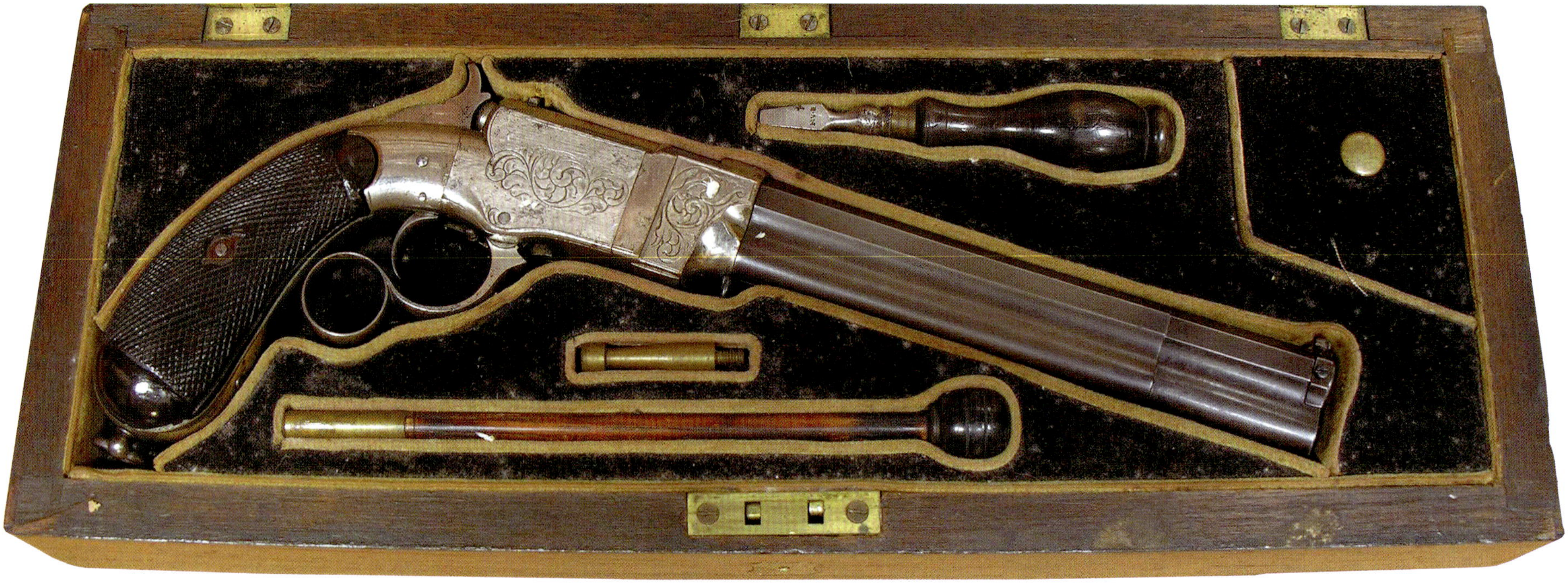

Cased Second Model Venditti Pistol. Roger Quisenberry collection

bibliography

BOOKS

Axelrod, Alan, *The Horrid Pit, The Battle of the Crater, The Civil War's Cruelest Mission*, Carroll & Graf Publishers, New York, 2007.

Bowman, Hank Wieand, *Famous Guns From the Winchester Collection*, Arco Publishing Company, Inc. New York, 1969.

Carmen, W.Y., *A History of Firearms from Earliest Times to 1914*, Dover Publications, Mineola, New York, 1955.

Catton, Bruce, *The Coming Fury: The Centennial History of the Civil War, Volume I*, Doubleday and Company Inc., New York, 1961.

Caulkins, Frances M., *The History of Norwich, Connecticut From Its Possession by the Indians to the Year 1866*, Friends of the Author, Norwich, Connecticut, 1874.

Cleveland, Horace W.S., *Hints to Riflemen*, D. Appleton & Co., New York, New York, 1864.

Cooper, Carolyn and Merrill Lindsay, *Eli Whitney and the Whitney Armory,* Eli Whitney Museum, Hamden, Connecticut, 1990.

Dana, Richard H., *To Cuba and Back, a Vacation Voyage*, Tichnor and Fields, Boston, 1859.

Edwards, William B., *Civil War Guns*, The Stackpole Co., Harrisburg, Pennsylvania, 1962.

Flayderman, Norm, *Flayderman's Guide to Antique American Firearms...and their values*, 8th Edition, Krause Publications, Iola, WI, 2001.

Giles, Ray T. and Daniel L. Shuey, *One Hundred Years of Winchester Cartridge Boxes 1856–1956*, Schiffer Publishing Ltd., 2006.

Haven, Charles T., "The History of the Winchester Repeating Rifle and The Other Arms and Ammunition made by The Winchester Repeating Arms Company and its Predecessors from 1850 to 1946," unpublished manuscript, 1947, Cody Firearms Museum.

Henshaw, Thomas, *The History of Winchester Firearms, 1866–1992*, Winchester Press, Linton, New Jersey, 1993.

Hoopes, Thomas T., *Arms and Armor Annual, Volume I*; edited by Robert Held, Digest Books, Inc., Northfield, Illinois, 1973.

Houze, Herbert G., *Winchester Repeating Arms Company, Its History & Development from 1865 to 1981*, Krause Publications, Iola, Wisconsin, 1994.

Jinks, Roy G., *History of Smith & Wesson*, Beinfield Publishing Inc., North Hollywood, California, 1977.

Krasnickas, Frank, *Cartridges of South Coventry*, Trafford Publishing, Victoria, B.C., 2006.

Lewis, Col. Berkeley R., *Notes On Ammunition of the American Civil War,* The American Ordnance Association, Washington, D.C., 1959.

Madis, George, *The Winchester Book*, Taylor Publishing Company, Dallas, Texas, 1971.

Madis, George, *The Winchester Book*, Art and Reference House, Brownsboro, Texas, 1985.

Malloy, John J., Dean S. Thomas and Terry A. White, *Percussion Ammunition Packets*, Thomas Publications, Gettysburg, Penn., 2003.

Manuscript in the McCracken Research Library, Buffalo Bill Historical Center, Cody, Wyoming.

Marcot, Roy M., *Civil War Chief of Sharpshooters, Hiram Berdan, Military Commander and Firearms Inventor*, Northwood Heritage Press, Irvine, California, 1989.

McDowell, R. Bruce, *Evolution of the Winchester*, Armory Publications, Tacoma, Washington, 1985.

Nolan, Dick, *Benjamin Franklin Butler, The Damnedest Yankee*, Presidio Press, Novato, California, 1991.

Olson, Dale, *Lever Action Magazine Rifles*, Taylor Publishing Company, Dallas, Texas, 1976.

Parsons, John E., *The First Winchester, The Story of the 1866 Repeating Rifle*, William Morrow and Company, New York, 1955.

Prime, William Cowper, *Boat Life in Egypt and Nubia*, Harper & Brothers, New York, 1857.

Prime, William Cowper, *Tent Life in the Holy Land*, Harper & Brothers, New York, 1857.

Quick, Les, *The Story of Benjamin Tyler Henry and His Famed Repeating Rifle*, Graphic Publishers, Santa Ana, California, 2008.

Rattenbury, Richard, "Winchester Promotional Arts: Graphic, Three-Dimensional and Live Advertising Mediums," Buffalo Bill Historical Center, Cody, Wyoming, 1978.

Richmond, John, *Egypt 1798–1952*, Columbia University Press, New York, 1977.

Sawyer, Charles Winthrop, *Our Rifles*, Boston, 1920.

Sellers, Frank M., *The William M. Locke Collection*, The Antique Armory, Inc., East Point, Georgia, 1973.

Supica, Jim and Richard Nahas, *Standard Catalog of Smith & Wesson*, Gun Digest Books, Iola, Wisconsin, 2006.

Thomas, Dean S., *Round Balls to Rimfire, Part Three,* Thomas Publications, Gettysburg, Pennsylvania, 2003.

Trudeau, Noah Andre, *The Last Citadel*, Petersburgh, Virginia, June 1864–April 1865, Louisiana State University Press, 1991.

Veader, D.H. and A.W. Earle, *The Story of the Winchester Repeating Arms Company*, unpublished manuscript, 1918, McCracken Research Library, Buffalo Bill Historical Center, Cody, Wyoming.

Warner, Ezra J., *Generals in Blue: Lives of the Union Commanders*, Louisiana State University Press, 1964.

Watrous, George, *The History of Winchester Firearms, 1866–1966*, Olin Mathieson Chemical Corporation, 1966.

Wilson, R.L., *Winchester, An American Legend*, Random House, New York, 1991.
Wilson, R L., *Winchester Engraving*, 2nd Edition, Beinfeld Books, Palm Springs, California, 1975.

JOURNALS & MAGAZINES

American Decorations, A List of the Awards of the Congressional Medal of Honor, the Distinguished-Service Cross and the Distinguished Service-Medal, Awarded Under the Authority of the Congress of the United States, 1862–1926, Adjutant General of the Army, 1927.
Burnside, Graham, "The Volcanic Quandry," *The Gun Report*, December 1958.
Chernoff, Arnold M., "The Volcanic Arms Story," *Guns & Ammo*, February 1962.
Gun Report Monograph, No. 9, "The Volcanic Repeating Rifle," The Gun Report Inc., Akron, Ohio, 1940.
Hall, Thomas E., "Forerunners of the First Winchester," *Gun Digest*, 12th Edition, Part 2 of 3, The Gun Digest Company, Chicago, Illinois, 1958.
Hall, Thomas E., "Forerunners of the First Winchester," *Gun Digest*, 13th Edition, Part 3 of 3, The Gun Digest Company, Chicago, Illinois, 1959.
Hall, Thomas E., "The Jennings and Volcanic," *Arms Gazette*, February 1974.
House, Kurt, "An Inventor's Pistol," *The Texas Gun Collector*, The Texas Gun Collector's Association, Spring 1991.
Houze, Herbert G., Journal, Smith & Wesson Collectors Association, Volume 41, Number 1, Spring 2007.
Houze, Herbert G., "The Henry and Model 1866 Serial Numbering," *Man at Arms*, July/August 1991.
Lewis, Col. B.R., "The Volcanic Arms," *American Rifleman*, November 1957.
Lewis, Col. B.R., "Inventors Won't Quit on the Rocket Ball," *American Rifleman*, July 1972.
Lewis M.D., Edmund, "The Volcanic Pistol Carbine," *The Winchester Collector*, Spring 2005.
Lewis M.D., Edmund, "Volcanic Firearms," *Gun Report*, September 2005.
Report of the Mechanics Institute of the City of San Francisco, San Francisco, California, 1858.
Satterlee, *Catalogue of Firearms for the Collector*, Detroit, 1940.
Serven, James E., "The History of the Volcanics," *Guns*, April 1965.
Serven, James E., "Early Lever Actions," *American Rifleman*, January 1967.
The Democratic Age Advertiser, R. Edwards & Company, New York, February 1859.
The Tribune Almanac, Greeley & McElrath, New York, 1857.

NEWSPAPERS

Daily Free Democrat, [Milwaukee], July 10, 1855.
Evening Bulletin, San Francisco, California, June 11, 1861, p. 1.
Frank Leslie's Illustrated Newspaper, October 9, 1858, p. 291.
Hartford Daily Courant, November 12, 1849, p. 2.
Independent American, Platteville, Wisconsin, April 2, 1855, p. 3.
Milwaukee Daily Sentinel, March 30, 1855, p. 4.
New Haven Palladium, March, 1855.
New Haven Register, January 26, 1891.
New York Herald, November 21, 1861, p. 5.
New York Times, Issues of February, March 12 and August 31, 1859.
New York Times, March 16, 1898.
Republican Compiler, Gettysburg, Pennsylvania, March 26, 1855, p. 1.
Scientific American, Vol. 12, Issue 12, November 29, 1856, p. 96.
Scientific American, Vol. 12, Issue 13, December 6, 1856, p. 101.
The Fond du Lac Union, Fond du Lac, Wisconsin, June 7, 1855, p. 2.
The Knickerbocker: New York Monthly Magazine, Vol. LIV, New York, 1859, p. 229.

OTHER

Carroll's New York City Directory, undated but probably late 1850s.
Compiled Records Showing Service of Military Units in Volunteer Union Organizations, Publications Number M594, 1965.
New Haven Arms Company Letter Press Book, McCracken Research Library, Buffalo Bill Historical Center, Cody, Wyoming.
The National Archives, Military and Service Pension Records, Washington, D.C.

endnotes

CHAPTER ONE

[1] Kane, Joseph Nathan, *Necessity's Child — The Story of Walter Hunt*, McFarland & Company, Inc., Jefferson, North Carolina, 1997.
[2] *Ibid.*, p. 139.
[3] *Ibid.*, p. 144.
[4] Henshaw, Thomas, *The History of Winchester Firearms 1866–1992*, Winchester Press, Linton, New Jersey, 1993, p. 1.
[5] Kane, Joseph Nathan, *op. cit.*, p. 73.
[6] *Ibid.*, p. 74.
[7] *Ibid.*, p. 75.
[8] *Ibid.*, pp. 146–148.
[9] *Ibid.*, p. 151.
[10] Hall, Thomas E., "Forerunners of the First Winchester: Part 2 of 3," *Gun Digest*, 12th edition, Gun Digest Company, Chicago, Illinois, 1958, p. 120.
[11] Kane, Joseph Nathan, *op. cit.*, p. 153.
[12] McDowell, R. Bruce, *Evolution of the Winchester*, Armory Publications, Tacoma, Washington, 1985, p. 3.
[13] *Ibid.*, p. 4–5.
[14] Kane, Joseph Nathan, *op. cit.*, p. 161.
[15] McDowell, R. Bruce, *op. cit.*, p. 3.
[16] *Ibid.*, p. 2.
[17] *Ibid.*, p. 4.
[18] Kane, Joseph Nathan, *op. cit.*, pp. 156–161.
[19] McDowell, R. Bruce, *Development of the Henry Cartridge and Self-contained Cartridges for the Toggle-Link Winchesters*, A.M.B., Metuchen, New Jersey, 1984, pp. 41–42.
[20] Carmen, W.Y., *A History of Firearms from Earliest Times to 1914*, Dover Publications, Mineola, New York, 1955, p. 163.
[21] McDowell, R. Bruce, *op. cit.*, p. 26.
[22] Houze, Herbert G., *Winchester Repeating Arms Company — Its History & Development from 1865 to 1981*, Krause Publications, Iola, Wisconsin, 1994, p. 10.
[23] Kane, Joseph Nathan, *op. cit.*, p. 271.
[24] McDowell, R. Bruce, *Evolution of the Winchester,* R. Bruce McDowell, Armory Publications, Tacoma, Washington, 1985, p. 11.
[25] Kane, Joseph Nathan, *op. cit.*, p. 161.
[26] McDowell, R. Bruce, *op. cit.*, p. 3.
[27] Quick, Les, *The Story of Benjamin Tyler Henry and His Famed Repeating Rifle*, Graphic Publishers, Santa Ana, California, 2008, p. 21.
[28] Henshaw, Thomas, *op. cit.*, p. 3.
[29] McDowell, R. Bruce, *op. cit.*, p. 11.
[30] *Ibid.*, p. 10.
[31] Henshaw, Thomas, *op. cit* p. 3.
[32] Houze, Herbert G., "The Jennings' Patent Rifle," *The Winchester Repeater*, Spring, 2009, p. 11.
[33] McDowell, R. Bruce, *op. cit.*, p. 10.
[34] *Ibid.*, p. 10.
[35] Flayderman, Norman E., *Flayderman's Guide to Antique American Firearms...and their values,* Gun Digest Books, Iola, Wisconsin, 2007, p. 302.
[36] McDowell, R. Bruce, *op. cit.*, p. 18.
[37] *Ibid.*, p 26.
[38] *Ibid.*, p. 29.
[39] Flayderman, Norman E., *op. cit.*, p. 302.
[40] McDowell, R. Bruce, *op. cit.*, p. 19.
[41] *Ibid.*, p. 29.
[42] Houze, Herbert G., *op. cit.*, pp. 10–11.
[43] Jinks, Roy G. and Sandra C. Krein, *Images of America: Smith & Wesson*, Arcadia Publishing, Charleston, South Carolina, 2006, p. 7.
[44] Jinks, Roy G., *History of Smith & Wesson*, Beinfield Publishing, Inc, North Hollywood, California, 1977, p. 2.
[45] McDowell, R. Bruce, *op. cit.*, p. 33.
[46] *Ibid.*, p. 18.
[47] Houze, Herbert G., *op. cit.*, p. 6.
[48] McDowell, R. Bruce, *op. cit.*, p. 57.
[49] Flayderman, Norman E., *op. cit.*, p. 302.
[50] *Ibid.*, pp. 302–303.
[51] *Ibid.*, p. 303.
[52] Madis, George, *The Winchester Book*, Art and Reference House, Brownsboro, Texas, 1985, p. 16.

CHAPTER TWO

[1] McDowell, R. Bruce, *Evolution of the Winchester*, Armory Publications, Tacoma, Washington, 1985, p. 60. Bruce McDowell's, book was the first to relate, in detail, the story of the Smith, Smith-Jennings, Smith & Wesson lever action firearms and those of the Volcanic Repeating Arms Company and the New Haven Arms Company. The authors refer the reader to his book for greater technical discussion of the design, function and performance of these firearms. The book goes into these matters in far greater detail.
[2] *Ibid.*, p. 84.
[3] Jinks, Roy G., *History of Smith & Wesson Firearms*, Beinfield Publications, Inc., North Hollywood, California, 1977, p. 6.
[4] Jinks, Roy G. and Sandra C. Krein, *Images of America: Smith & Wesson*, Arcadia Publications, Mt. Pleasant, South Carolina, 2006, p. 7.
[5] Jinks, Roy G., *History of Smith & Wesson Firearms*, p. 13.
[6] McDowell, *op. cit.*, p. 59.
[7] Edwards, William B., *Civil War Guns*, The Stackpole Company, Harrisburg, Pennsylvania, 1962, p. 165.
[8] Jinks, *op. cit.*, p. 20.
[9] McDowell, *op. cit.*, p. 60.
[10] Sellers, Frank, "The Venditti Myth," *Smith & Wesson Collector's News*, March and June, 1973, pp. 87–91.
[11] Thomas, Dean S., *Round Ball to Rimfire*, Part Three, Thomas Publications, Gettysburg, Pennsylvania, 2003, p. 217.
[12] Quick, Les, *The Story of Benjamin Tyler Henry and His Famed Repeating Rifle*, Graphic Publishers, Santa Ana, California, 2008, p. 21.
[13] Lewis, Col. Berkeley R., "The Volcanic Arms and Ammunition," *American Rifleman*, November, 1957, p. 49
[14] Parsons, John E., *The First Winchester*, Winchester Press, New York, New York, 1969, p. 8.
[15] Houze, Herbert G., *Winchester Repeating Arms Company — Its History & Development from 1865 to 1981*, Krause Publications, Iola, Wisconsin, 1994, p. 13.
[16] McDowell, R. Bruce, *op. cit.*, pp. 31–32.
[17] Lewis, Col. B.R., *op. cit.*, p. 48.
[18] McDowell, R. Bruce, *op. cit.*, pp. 67–68.
[19] Jinks, Roy, *op. cit.*, p. 23.
[20] *Ibid.*, p. 32.
[21] Jinks, Roy, *op. cit.*, p. 25.
[22] *Independent American*, Platteville, Wisconsin, April 2, 1855, p. 3.
[23] *Milwaukee Daily Free Democrat*, May 26, 1855.
[24] Jinks, Roy, *op. cit.*, p. 25.
[25] Suydam, C.R., "A Cased Venditti Pistol," *The Gun Report*, December, 1972, p. 58.
[26] Flayderman, Norm, *Flayderman's Guide to Antique American Firearms*, 8th Ed., Krause Publications, Iola Wisconsin, 2001, p. 264.
[27] Walter Hallstein, personal communication.
[28] McDowell, R. Bruce, *Development of the Henry Cartridge*, J&B Flood, Edison, New Jersey, p. 34. McDowell's monograph contains detailed descriptions and discussion of the early development of metallic cartridges including the Pauly self-primed center-fire cartridge of 1812, the Galy-Cazalat centerfire cartridge of 1826 and similar cartridges by William Bush and Pottet in 1840. Smith and Wesson were latecomers to the game.
[29] Jinks, Roy, *op. cit.*, p. 32.
[30] *Ibid.*, pp. 31–32.
[31] McDowell, R. Bruce, *Evolution of the Winchester*, Armory Publications, Tacoma, Washington, 1985, p. 72.
[32] Charles May, personal communication, Cody Firearms Museum, Buffalo Bill Historical Center, Cody, Wyoming.
[33] McDowell, R. Bruce, *op. cit.*, p. 72.
[34] Charles May, personal communication.
[35] McDowell, R. Bruce, *op. cit.*, p. 76.
[36] *Ibid.*, p. 70.
[37] *Ibid.*, p. 76.
[38] *Ibid.*
[39] *Ibid.*
[40] Jinks, Roy, *op. cit.*, p. 31.
[41] Roy Jinks, personal communication.
[42] Jinks, Roy, *op. cit.*, p. 31.
[43] McDowell, R. Bruce, *op. cit.*, p. 78.
[44] *Ibid.*, p. 78.
[45] Hall, Thomas E., "Forerunners of the 1st Winchester," *Gun Digest*, Pt. II, 12th Ed., The Gun Digest Company, Chicago, Illinois, 1958, p. 124.
[46] Roy Jinks, personal communication.
[47] Gerald Klaz, M.D., personal communication.
[48] McDowell, R. Bruce, *op. cit.*, p. 80.
[49] *Ibid.*, p. 80.
[50] Hall, Thomas E., *op. cit.*, p. 124.
[51] McDowell, R. Bruce, *op. cit.*, p. 84.
[52] Jinks, Roy, *op. cit.*, pp. 34–36.
[53] Lewis, Col. B.R., *op. cit.*, p. 49.
[54] Supica, Jim and Richard Nahas, *Standard Catalog of Smith & Wesson*, Gun Digest Books, 2006, p. 65.

CHAPTER THREE

1 Krasnickas, Frank, *Cartridges of South Coventry*, Trafford Publishing, Victoria, B.C., 2006. An exhaustive study of the many South Coventry ammunition manufacturers of the period, it includes biographies of the key players involved in the fabrication of these cartridges, their patents and illustrations of the cartridges and the cartridge boxes as well as various other manufacturers of the period.
2 *Ibid.*, p. 19.
3 *Ibid.*, p. 3.
4 *Ibid.*, p. 2.
5 *Ibid.*, p. 10.
6 McDowell, R. Bruce, *Development of the Henry Cartridge*, J&B Flood, Edison, New Jersey, 1984, p. 16.
7 *Ibid.*, p. 15.
8 McDowell, R. Bruce, *Evolution of the Winchester*, Armory Publications, Tacoma, Washington, 1985, p. 2.
McDowell also asks whether the gun or cartridge came first and relates that, "On September 29, 1812,... Jean Samuel Pauly was granted a French patent to fire a self-contained center fire cartridge, having an internal primer and a rim at the base of the casing. This was the first gun to use any form of self-contained center fire cartridge with percussion ignition contained within the base of the casing. Since the Pauly cartridge was invented before the percussion cap, all such cartridges used a priming (fulminate) powder that was tamped into the priming receptacle at the base of the cartridge... The Pauly center fire cartridges were designed for pistols, rifles and shotguns. A major problem that faced the early gunmakers, including Hunt, Jennings and Smith and Wesson, was that of escaping gases, resulting in greatly reduced fire power. Pauley's cartridge bases were formed with a tapered rim. This tapered rim would form a gas tight seal when placed within the tapered chamber of the gun barrel, and would also prevent the cartridge from going too far into the barrel itself. Immediately behind the tapered rim, was an extraction groove, such as found on the later Pottet cartridge. This obviously was the concept later used in many metallic cartridges."
9 *Ibid.*, p. 10.
10 Jinks, Roy G., *op. cit.*, p. 20. Jinks relates that these comments were from the D.B. Wesson Daybook.
11 *Ibid.*
12 Serven, James E., *Guns*, April 1965, p. 33.
13 McDowell, R. Bruce, *op. cit.*, p. 29.
14 Hall, Thomas E., *Forerunners of the First Winchester*, Part 2 of 3, The Gun Digest Company, 12th Edition, 1958, p. 127.
15 Burnside, Graham, "The Volcanic Quandry," *The Gun Report*, December 1958, p. 37.
16 *Ibid.*, p. 37.
17 *Ibid.*
18 Lewis, Col. B.R., "The Volcanic Arms," *American Rifleman*, November 1957, p. 51. Colonel Berkeley R. Lewis (Retired) was in the Ordnance Corps of the United States Army and the author of "Small Arms and Ammunition in the United States Service."
19 *Ibid.*, p. 51.
20 Burnside, *op. cit.*, p. 37.
21 Kelley, Alan, personal communication.
22 Veader, D.H. and A.W. Earle, *The Story of the Winchester Repeating Arms Company*, privately printed, 1918, p. 8.
The authors were longstanding employees of the Winchester Repeating Arms Company. The information is found in a typed manuscript in the McCracken Library of the Cody Firearms Museum. There are of four known copies. Two copies were placed in the Gun Museum in New Haven, Connecticut, by then curator Thomas E. Hall. Two other copies are in private hands. *The Story of the Winchester Repeating Arms Company* contains a large amount of information on the predecessors of Winchester, including the Smith & Wesson Company, the Volcanic Repeating Arms Company and the New Haven Arms Company. It includes a great amount of detail on the formation of these companies as well as their stockholders, company officers, employees and their business dealings, including patent disputes and lawsuits. Usually overlooked on pp. 24–25 of *The Winchester Book* by George Madis is a manuscript signed by Daniel H. Veader on September 26, 1866. It illustrates the difficulties collecting royalties from the Union Metallic Cartridge Company on the cartridge patents owned by the New Haven Arms Company.
23 *Ibid.*, p. 8.
24 *Ibid.*
25 Krasnickas, *op. cit.*, p. 29.
26 Lewis, Col. B.R., "Inventors Won't Quit on the Rocket Ball," *American Rifleman*, July 1972, p. 39.
27 *Ibid.*, p. 39.
28 See Appendix II on page 142 of this book.
29 Herb Houze relates that Jared Wilson Post was living in Westbrook, Connecticut, in 1850 and was listed in the census of that year as "agent." In 1856, he was listed in Trow's New York City Directory for 1857 as "Post. J.W., firearms." Veader and Earle, *op. cit.*, p. 19, relates that he was succeeded by Joseph Merwin as agent for the New Haven Arms Company in New York until 1859.
30 Thomas, Dean S., *Round Ball to Rimfire*, Part III, Thomas Publications, Gettysburg, Pennsylvania, 2003.
31 Lewis, Col. B.R., *op. cit.*, p. 50.
32 Klaz, M.D., Gerald, personal communication.
33 Kass, George, personal communication.
34 New Haven Arms Company Letterpress Book, *op. cit.*, p. 115.

CHAPTER FOUR

1 Veader, D.H. and A.W. Earle, *The Story of the Winchester Repeating Arms Company*, privately printed, 1918, p. 1. D.H. Veader entered the employ of the Winchester Repeating Arms Company in 1866 and A.W. Earle joined the firm on August 11, 1883.
They listed these subscribers to the stock:
Paschal Converse, 41 St. John St., New Haven (clock maker)
Charles B. Bates
James N. Adams, 41 St. John St., New Haven (clock maker)
C.B. Bryant, 41 St. John St., New Haven (clock maker)
Morris Tyler, 77 George St., New Haven (shoe business)
E.B. Bowditch, 117 Orange St., New Haven (cabinet maker)
Philip Pond, 41 John St., New Haven (clock maker)
James R. Campbell
John E. Alcott, 95 Wallace St., New Haven (joiner)
I.C. Stock, 68 Wooster St., New Haven
(railroad conductor, N.Y., N.H. & H.R.R.) Green Kendrick
Oliver F. Winchester, 57 Court St., New Haven (shirt manufacturer)
Nelson H. Gaston, 137 Orange St., New Haven (shipping merchant)
Boo Lewis
Charles Dickerman, Howe & George Sts., New Haven
(coach manufacturer)
William C. Hicks, New York
Silas R. Sperry, 84 Hamilton St., New Haven (clock maker)
Henry Hooker, 96 York St., New Haven (carriage manufacturer)
R.N. Buchanan
Ruben Stannard
George W. Howell
Hiram Camp, 76 Wooster St., New Haven (President N.H. Clock Co.)
Sackett Gilbert, 43 Crown St., New Haven (saddle manufacturer)
H.T. & C.S. Bushnell, Grocers, State & Crown Sts. New Haven
(C.S. Bushnell was later one of the promoters of the Monitor)
James N. Linsley
Samuel L. Talcott
J.W. Post
George Bishop, 37 Hamilton St., New Haven (carriage maker)
William Strong
William Hustace
Charles S. Harris
Thaddeus Sherman, 20 High St., New Haven
James B. Campbell
William H. Robertson

It should be noted that D.H. Veader and A.W. Earle were employed by the Winchester Repeating Arms Company in May 1869 and August 1883, some sixty years after the incorporation of the Volcanic Repeating Arms Co. It is assumed that they had access to the appropriate records. However, it has been said by some historians that some of their information should be taken with a grain of salt. The following interesting inventory, as of January 1, 1860, is taken from their manuscript:

Finished, 164 carbines, 971 pistols list $21,020.25 – 50%	$10,510.13
Ready for proving, 23 carbines, 178 pistols list 3, 001.00 – 50%	1,500.50
Partly finished, 220 carbines	3,152.50
Partly finished, 500 pistols	2,375.00
Component parts for "old style" pistols	126.85
Parts in process	2,541.70
Volcanic ammunition list 1,727.00 – 50%	863.50
Raw materials	2,685.67
Goods in New York, 40 carbines, 312 pistols	3,212.63
Volcanic ammunition	214.00
Goods in transit, 13 carbines, 44 pistols list 1,179.80 – 50%	589.90
Goods of other makes	445.75
Total	**$28,218.13**

2 *Ibid.*, p. 4.
3 *Ibid.*
4 *Ibid.*, p. 5.
Jared Wilson Post, born in 1815, in 1850 was living in Westbrook, Connecticut and listed in the 1850 census as "Agent." In 1856, Trow's Directory, New York City for 1857 listed him as "Post, Jared W., firearms, 23 John, h New Haven, Conn."
5 McDowell, R. Bruce, *Evolution of the Winchester*, Armory Publications, Tacoma, Washington, 1985, p. 84.
6 Madis, George, *The Winchester Book*, Taylor Publishing Company, Dallas, Texas, 1991, p. 19.
7 McDowell, R. Bruce, *op. cit.*, p. 84.
8 *Ibid.*, p. 84.
9 Serven, James E., "The History of the Volcanics," *Guns*, April 1965, p. 41.
10 Houze, Herbert G., *Winchester Repeating Arms Company — Its History & Development from 1865 to 1981*, Krause Publications, Iola,

Wisconsin, 1994, p. 8.
[11] McDowell, R. Bruce, *op. cit.*, p. 84.
[12] *Ibid.*, p. 99.
[13] *Ibid.*, p. 72.
[14] *Ibid.*, p. 110.
[15] Serven, James E., "Early Lever Actions," *American Rifleman*, January 1967, p. 34.
[16] McDowell, R. Bruce, *op. cit.*, p. 110.
[17] *Ibid.*, p. 110.
[18] Hall, Thomas E., "Forerunners of the First Winchester," *Gun Digest*, Part II, The Gun Digest Company, 13th Edition, Chicago, Illinois, 1959, p. 126.
[19] Jinks, Roy, *op. cit.*, p. 34. Rollin White, an ex-Colt employee, had patented a revolving-cylinder gun in which the cylinder had been bored end to end which allowed cartridges to be loaded from the rear. His U.S. Patent No. 12,649 was one of the most significant in the development of the revolver allowing metallic cartridges to be used rather than the cap and ball system.
[20] Jinks, Roy E., *op. cit.*, p. 36.

CHAPTER FIVE

[1] McDowell, R. Bruce, *Evolution of the Winchester*, Armory Publications, Tacoma, Washington, 1985, p. 99.
[2] *Ibid.*, p. 99.
[3] *Frank Leslie's Illustrated Newspaper*, October 9, 1858, p. 291.
[4] *Ibid.*, p. 102.
[5] *Ibid.*
[6] *Ibid.*
[7] *Ibid.*
[8] *Ibid.*, p. 105.
[9] *Ibid.*, p. 108.
[10] *Ibid.*, p. 110.
[11] *Ibid.*, p. 107.
[12] *Ibid.*
[13] *Ibid.*, p. 106.
[14] *Ibid.*, p. 107.
[15] *Ibid.*, p. 110.
[16] Letter from George Maddis.
[17] *Ibid.*, p. 110.
[18] Hall, Thomas E., "Forerunners of the First Winchester," *Gun Digest*, Part II, The Gun Digest Company, 13th Ed., Chicago, Illinois, 1959, p. 127.

Dimensions of "Volcanic" Barrels and Bullets
Smith & Wesson Pistols
4-inch barrel, bore diameter .302". 8-inch barrel, bore diameter .386".
Volcanic Pistols
In eight pistols having barrels of 6" or 8" long, all had bore diameters of .387".
New Haven Pistols
Five with 3½-inch barrels (referred to in contemporary advertising as 4-inch barrels) had bore diameters of , 300", .301", .302", .304".
One 6" barrel measured .300".
New Haven Rifles
Seven barrels (16", 20" and 24" were checked and all but one 24" barrel had bore diameters of .393". The one exception measured .396". One 16" barrel had a bore diameter of .393" and a groove diameter of .405". An 8" barrel pistol checked the same way measured .387" bore and .405" groove. None of the cartridges measured were perfectly round; the diameters given are an average of the varying dimensions found. One cartridge measured .433" and weighed 124 grains. Four others of the larger Volcanic cartridges were measured and found to be as follows: Average diameter at largest point .411", .325", .326", .327"

[19] McDowell, R. Bruce, *op. cit.*, p. 109.
[20] Thomas E. Hall personal communication.
[21] McDowell, R. Bruce, *op. cit.*

CHAPTER SIX

[1] Herbert G. Houze, personal communication.
Richard Bates Inshaw (1805–1865) was a gunsmith and engraver living in Springfield, Massachusetts, and was listed in Greer's Express Directory and Railway Forwarders Guide, Springfield. He worked for the Ames Company through 1851 and decorated a number of presentation swords, including one for General Winfield Scott. He later engraved pistols for the Massachusetts Arms Company and other firearms firms.
[2] Thomas E. Hall, personal communication.
[3] Bruce Jordan, Donovan Williamson, personal communications. Both of these collectors have kept detailed notes for many years on the number of 25-inch carbines that have been offered, either by private treaty or through auction. Williamson has kept auction records for more than twenty years and found only a few offered.
[4] Thomas E. Hall, personal communication.

CHAPTER SEVEN

[1] House, Kurt, "An Inventor's Pistol," *The Texas Gun Collector*, The Texas Gun Collector's Association, Spring, 1991, p. 48.
[2] *Ibid.*, p. 52.
[3] Wilson, R.L., *Winchester Engraving*, Beinfield Books, Palm Springs, California, 1975, p. 9.
[4] Richmond, John, *Egypt 1798–1952*, Columbia University Press, New York, 1977, p. 257.
[5] Marcot, Roy M., *Civil War Chief of Sharpshooters*, Hiram Berdan, Military Commander and Firearms Inventor, Northwood Heritage Press, Irvine, California, p. 101.
[6] Letter, War Department, Inspector General's Office, Washington, D.C., July 27, 1893.
[7] Albee…American Decorations, A List of Awards of the Congressional Medal of Honor the Distinguished-Service Cross and the Distinguished-Service Medal,…1862–1926, Adjutant General of the Army, 1927.
[8] Flayderman, Norm, personal files and communication.
[9] Rutter, M.D., Stephen, "An Unusual 'Volcanic' Carbine Serial Number," *The Gun Report*, April 2009, pp. 14–18.
[10] Martin, Edward, *Always Be On Time — An Autobiography*, The Telegraph Press, Harrisburg, Pennsylvania, 1959, p. 87.
[11] Madis, George, *The Winchester Book*, Art and Reference House, Brownsboro, Texas, 1985, p. 25.

CHAPTER EIGHT

[1] National Archives, Military & Service Pension Records, Washington, D.C.
[2] *Ibid.*
[3] Alden S. Bradstock, III is a military historian and served as a Field Artillery officer in the US Army for eleven years.
[4] Axelrod, Alan, *The Horrid Pit, The Battle of the Crater, The Civil War's Cruelest Mission*, Carroll & Graf Publishers, New York, 2007.
[5] Nolan, Dick, *Benjamin Franklin Butler, The Damnedest Yankee*, Presidio Press, Novato, California, 1991, p. 280.
[6] *Ibid.*
[7] Houze, Herbert G., "The Volcanic in Service During 'The Staten Island War' of 1858," *Man at Arms*, October 2005, pp. 25–31.
[8] Dana, Richard H., *To Cuba and Back, A Vacation Voyage*, Tichnor & Fields, Boston, 1859.
[9] *The New York Times*, March 16, 1898.
[10] Prime, William Cowper, *Boat Life in Egypt and Nubia*, Harper and Brothers, New York, 1857, p. 495.
[11] *Ibid.*, p. 196.
[12] *Ibid.*, p. 254.
[13] *Ibid.*, pp. 414–415.

CHAPTER NINE

[1] Jinks, Roy, personal communication.
[2] Haven, Charles T., *The History of the Winchester Repeating Rifle and the Other Arms and Ammunition Made by The Winchester Repeating Arms Company and its Predecessors from 1850 to 1946*, unpublished work, Keller Firearms Research Library, Chapter 1, p. 14.
[3] Rattenbury, Richard, *Winchester Promotional Arts*, Buffalo Bill Historical Center, 1978, in Preface by George Madis.
[4] *Ibid.*, p. 3.
[5] McDowell, R. Bruce, *Evolution of the Winchester*, Armory Publications, Tacoma, Washington, 1985, p. 100.

CHAPTER TEN

[1] McDowell, R. Bruce, *Evolution of the Winchester*, Armory Publications, Tacoma, Washington, 1985, p. 124.
[2] McDowell, R. Bruce, *Development of the Henry Cartridge*, J&B Flood, Edison, New Jersey, 1984, p. 34.
[3] Houze, Herbert G., "John Parker Lindsay and John Walch: References and Biographical Notes," *The Gun Report*, May 2008, p. 24.
[4] McDowell, R. Bruce, *op. cit.*, p. 35.
[5] McDowell, R. Bruce, *Evolution of the Winchester*, p. 115.
[6] *Ibid.*, p. 116.
[7] Flayderman, Norm, *Flayderman's Guide to Antique American Firearms*, 8th Edition Krause Publications, Iola, Wisconsin, 2001, p. 306. There are several theories, and some controversy, among different authors regarding production figures and dates of production. For a complete discussion, the reader is referred to books on the Henry rifle by George Madis, Les Quick and Wiley Sword that are listed in the bibliography.
[8] *Ibid.*, p. 266.
[9] Houze, Herbert G., *Winchester Repeating Arms Company — Its History & Development from 1865 to 1981*, Krause Publications, Iola, Wisconsin, 1994, pp. 23–25.
[10] Les Quick, *op. cit.*, p. 23.
[11] Sword, Wiley, *The Historic Henry Rifle*, Andrew Mowbray Publishers, Lincoln, Rhode Island, 2002, p. 85.
[12] Notes taken by Kurt House from sources at the Cody Firearms Museum, at the Buffalo Bill Historical Center in 1980, reveal interesting information about B. Tyler Henry, the man. These were transcribed from a small pocket notebook that was dated May 10, 1904. R.L. Henry, his nephew, said that Uncle "Tylah" was moody and given to fits of temper, was frequently withdrawn and "temperamental." According to his nephew, Henry's temperament was such that as a machinist, he might become disgusted when a part did not come out right and walk off for the rest of the day. Thomas E. Hall, then

curator of the Winchester Gun Museum in New Haven, also related to House that Hall's uncle, when a small boy, worked as a machinist's helper for Henry in his barn in Claremont, New Hampshire. Also, the uncle told Hall that Henry and his wife lived together for several years without speaking to each other. He had five daughters who, perhaps, contributed to the problem. House was also told by Hall that the powder horn in the Cody Firearms Museum was made by Henry when he was a small boy. Interesting as well is that the Rimfire Cartridge Board in the Cody Firearms Museum, the only authentic one known, was brought into the Winchester Company by a fellow who cleaned barns in New Haven.

13 Parsons, John E., *The First Winchester*, William Morrow & Company, New York, 1955, pp. 50–51.

14 Randy Hughes, personal communication.
Randy Hughes has made an extended study of the 1st Model 1866 rifle. He has kept copious records and thinks that fewer than ten were produced. More 1st Model carbines were manufactured than rifles. Several well-made reproductions of the 1st Model rifle are known.

APPENDIX III

1 Marcianò, Emanuele, "The Final Word on the Venditti-Volcanic Controversy," *Arms and Armor Annual*, Volume I, Robert Held, editor, Digest Books, Inc., Northfield, Illinois, 1973, p. 311.

2 Terenzi, Marcello, "The Italian Origin of the Winchester," paper presented at conference for American Society of Arms Collectors, New Orleans (Missouri), 1970.

3 Marcianò, Emanuele, *op. cit.*, p. 309.

4 *Ibid.*, p. 312.

5 McDowell, R. Bruce, *Evolution of the Winchester*, Armory Publications, Tacoma, Washington, 1985, p. 187.

6 Marcianò, Emanuele, *op. cit.*, p. 310.

7 McDowell, R. Bruce, *op. cit.*, p. 193.

8 Marcianò, Emanuele, *op. cit.*, p. 311.

9 McDowell, R. Bruce, *op. cit.*, p. 188.

10 *Ibid.*, p. 188.

11 Sellers, Frank, "The Venditti Myth," *Smith & Wesson Collector's News*, March and June 1973. p. 89.

12 *Ibid.*, p. 91.

13 *Ibid.*, p. 87.

14 Burnside, Graham, "The Volcanic Quandry," [sic] *The Gun Report*, 1958.

15 *Ibid.*, p. 7.

16 Sellers, Frank, *op. cit.*, p. 89.

17 *Ibid.*, pp. 91–92.

18 Appendix I, U.S. Patent No. 10,535.

19 Terenzi, Marcello, "The Italian Origin of the Winchester," American Society of Arms Collectors, fall 1970.

20 Sellers, Frank, *op. cit.*, p. 90.

21 *Ibid.*, p. 91.

22 *Ibid.*

23 *Ibid.*, p. 89.

24 *Ibid.*, p. 90.

25 Suydam, Charles R., "A Cased Venditti Pistol," *The Gun Report*, December 1972, pp. 58–59.

26 Marciano, Emanuele, *op. cit.*, pp. 308–314.

27 *Ibid.*, p. 309.

28 McDowell, R. Bruce, *op cit.*, p. 188.

29 Marciano, Emanuele, *op. cit.*, p. 314.

30 McDowell, R. Bruce, *op. cit.*, p. 196.

31 *Ibid.*, p. 194.

32 *Ibid.*, p. 188.

33 *Ibid.*, p. 191.

34 *Ibid.*, p. 195.

35 *Ibid.*, p. 193.

36 *Ibid.*, p. 195.

37 *Ibid.*, p. 188.

38 Marciano, Emanuele, *op. cit.*, p. 310.

39 *Ibid.*, p. 311.

40 McDowell, R. Bruce, *op. cit.*, p. 193.

41 *Ibid.*, p. 193.

42 Marciano, Emanuele, *op. cit.*, p. 313.

43 McDowell, R. Bruce, *op. cit.*, pp. 186–196.

44 *Ibid.*, p. 75.

45 *Ibid.*, p.193.

46 Marciano, Emanuele, *op. cit.*, p. 314.

47 McDowell, R. Bruce, *Development of the Henry Cartridge*, J&B Flood, Edison, New Jersey, p. 32.

48 Marciano, Emanuele, *op. cit.*, p. 314.

photo credits

Ray Bentley: p. 48 center left; p. 49 top left; pp. 71, 77 right; p. 81 top; p. 82 bottom; p. 91 bottom and right
Alden S. Bradstock III: pp. 23, 26, 27 left; p. 48 top; p. 55 top; pp. 56, 60 center and bottom; pp. 108, 109 right
Cody Firearms Museum: pp. 10, 40, 48 bottom right; p. 82 top; p. 93
Czerny's International Auction House: p. 145
David Condon: p. 74 bottom
Peter DeRose: p. 29 top; pp. 32, 37
George Mason University Special Collections: p. 110
Frank Graves: p. 105 right
Perry Hansen: p. 85
Herbert G. Houze: p. 52 top right; pp. 53, 128 top; p. 124
Randy Hughes: pp. 132–134
Roy G. Jinks: pp. 16, 19, 20 left; p. 23 top; p. 24 right; p. 25 left center; p. 45 center and bottom; pp. 66, 116 left
James D. Julia Inc.: p. 25 bottom right; p. 29 center; p. 64 bottom; pp. 89, 90 bottom; p. 92 bottom; pp. 103, 107 right
Rob Kassab: p. 148
Frank Krasnickas: p. 17 bottom left; p. 45 top; p. 49
Edmund Lewis: Frontispiece, pp. 59, 60 top; p. 63 bottom; pp. 64, 105 top left; p. 115 top left; p. 117 right; p. 118 right; pp. 120–124; p. 125 left; p. 126
Little John's Auctions: p. 105 bottom left
John Malloy: p. 43 top left
Greg Martin Auctions: p. 65
R. Bruce McDowell: p. 11 left and center; p. 29 bottom
Chip Morton: Cover, pp. 13, 14; p. 17 top and center; pp. 18, 24 left; p. 25 left and bottom; p. 27 center and right; p. 28 top and right; p. 33 left top, bottom and right; pp. 34, 36, 38, 39, 42; p. 43 bottom left; p. 54 bottom; p. 57 bottom; p. 58 right and bottom; p. 73 top; pp. 75, 77 left; p. 81 bottom; p. 84 bottom; pp. 86, 87, 90 top; p. 92 top; p. 94 bottom; pp. 95, 96, 98, 102, 106 right; p. 107 left; pp. 111, 112, 115 bottom; pp. 129, 130, 146
Stuart Mowbray: p. 106 left
Al Muchka: pp. 41, 61
Les Quick: p. 21 bottom; p. 127
Rock Island Auction Company: p. 74 top; p. 84 top, 88, 91 top left; p. 94 top; p. 131
Stephen Rutter: p. 15
Stuart Smith: p. 28 bottom left; pp. 31, 33 center left; p. 43 top; pp. 46, 54 top; p. 55 center and bottom; p. 57 top; pp. 62, 63, 69, 70, 72, 73 bottom; pp. 76, 78–80, 99, 100, 101, 104, 115, 126 right
Smithsonian Institution: p. 21, top
Dean Thomas: p. 48
Harold F. Williamson: pp. 51, 52, 129
R.L. Wilson: pp. 97, 118 left

index

A

Albee, George E., 99
Allen & Thurber, 16, 20, 31
Arrowsmith, George, 9–13, 15, 20

B

Babcock, M.L., 47,
Barrels:
  3-inch, 42
  3½-inch, 76, 78, 85, 90
  4-inch, 68, 71, 73, 83
  6-inch, 25, 28, 29, 42, 53–55, 58, 68, 69, 71, 73, 76–78, 98, 101
  6⅗-inch, 98
  8-inch, 25, 28, 50, 53–55, 58, 59, 61, 65, 68, 71, 73, 76, 78, 85, 88
  9½-inch, 98
  16-inch, 25, 53, 58, 59, 61, 62, 64, 65, 68, 70–73, 79, 81–84, 86, 92, 94
  20-inch, 68, 79, 85
  21-inch, 62, 64, 79, 81, 83, 86, 102
  24-inch, 62, 68, 79, 83
  25-inch, 58, 62–64, 68, 79, 80, 81, 83, 85 86, 92, 93
  Ribbed
  Type I, 63, 83, 84
  Type II, 79, 83, 102
  Type III, 68, 80, 81, 83, 84
Bennett, Thomas, G., 96
Breechloader, 11, 15, 16, 44, 122
Bullet lubrication (grease grooves), 15, 44, 46
Butler, Major General Benjamin Franklin, 109–111

C

Carrera, Gen., 96
Carrier, 10–12, 14–17, 24, 25, 29, 32, 34, 49, 61, 68, 76, 81, 89, 131
Cartridges:
  Centerfire, 12, 50
  Henry cartridge, .44 rimfire, 24, 47, 50, 127, 128
  Hunt Rocket Ball, .54 caliber, 9–13, 15, 17, 19, 22, 24, 43–46, 49
  Hunt Wooden cartridge, 12
  Metallic cartridge, 44–45
  Smith & Wesson .22 rimfire, 66
  Volcanic No. 1 Pistol Cartridge, .31 caliber, 45, 47, 49, 61
  Volcanic No. 2 Pistol Cartridge, .41 caliber, 45-47, 49, 61
Cody, William F. (Buffalo Bill), 99
Colt, Samuel, 9, 19, 43, 95, 116
Colt's, 22, 23, 47, 52, 54, 58, 59, 95, 105, 113, 114, 123
Converse, 96
Crittenden, Ralph, 43
Crittenden & Tibbals, 20, 43–49

D

Dixon, C.P., 14–15, 17, 118

E

Egypt, 98

F

Farouk, King, 98
Flobert cartridge, 20, 21, 24, 34, 44, 66
Flobert, Louis Nicholas Auguste, 20, 44

G

Gardner, Lt. Thomas W., 107
Gettysburg, 105
Grant, Ulysses S., 111
Great Exhibition, 20, 147

H

Henry, Benjamin Tyler, 15, 16, 18–21, 47, 66–68, 96, 103, 127, 128, 130–132
Henry rifle, 9, 21, 26, 40, 50, 67, 84, 95, 96, 102, 127–131
Hicks, William C., 21, 34, 52, 58
Hoggson, Samuel J., 95
Hotchkiss Repeating Rifle, 95, 99
Howard, Thomas, 105
Howe, Elias, 10
Hunt, Walter, 1–5, 7–10, 20–22, 24, 43–47, 49, 52

J

Jennings cleaning rod, 15
Jennings, Lewis, 11–13, 15, 16, 20, 31, 44
Jennings Rifle, 9, 11–22, 26, 44, 118

K

Keller, Charles M., 11
King, Nelson H., 132, 133
Kingsbury Box Company, 49, 65

L

Lawrence, Richard, 15, 20, 44
Leonard Pistol Works, 16, 19
Lewis, Col. B.R., 21
Lillie, Gordon (Pawnee Bill), 99

M
Martin, Edward, 102
Merwin, Joseph, 68, 112, 113, 119, 120, 123
Mexican War, 10-12, 19
Mt. Vesuvius, 23

N
Nazro, H.J., 23, 122
New Haven Arms Company, 12, 20–22, 46, 47, 49, 50, 53, 52–64, 66–70, 73, 78, 9, 83–86, 90, 92, 95, 98, 101, 102, 109, 112, 113, 118–120, 123, 124, 127, 128, 130–132
Nimschke, L.D., 95

P
Palmer, Courtlandt, 15–17, 19–22, 44, 51–53
Palmer, William, 12
Patents, 136
Pill primer, 10, 11, 13, 15, 17, 44–46
Pistol-Carbine, 36, 36, 37, 39, 53, 58–61, 63, 64, 68, 70, 71, 81, 83–85, 89, 90, 103
Pistols, large-frame, 16, 22, 23, 25–33, 35, 36, 45, 46, 54, 59, 71, 73, 78, 89, 96, 118, 119, 124
Pistols, small-frame, 16, 22, 23, 25, 26, 29–33, 35, 45, 46, 52, 71, 73, 78
Pistols:
- Type I, 24–26, 28–33, 86, 89
- Type II, 25, 26, 28–31
- Type III, 25, 26, 29, 30, 35, 36

Post, J.W., 22, 49, 53, 59, 62 68, 118, 123
Prime, William Cowper, 113–116

R
Ripley, J.B., 21
Robbins & Lawrence Company, 15–17, 20, 21, 25, 31, 44
Rocket Ball: see Cartridges
Rockwood, Theodore H., 109, 110

S
Smith, Horace, 9, 16, 18, 44, 51, 52, 66
Smith-Jennings Rifle, 9, 16–19, 21, 22, 26, 43, 44, 52
Smith & Wesson Company, 21, 22, 34, 40, 41, 45, 51, 95
Smith & Wesson Lever Action Firearms, 19–23, 25–42, 43–46, 52, 53, 58, 71, 73, 83, 118–126
Smith & Wesson Revolver Factory, 41, 66
Smith & Wesson Volcanic Ammunition, 23, 24, 43, 45, 47–50, 52, 53, 67, 68, 103, 113, 114, 116, 117, 120–122, 127, 128
Staten Island War, 111–113

T
Thimbling, 47
Thomas, Col. Henry G., 109
Tibbals, William, 43
Tiffany & Co., 96
Toggle-link, 21–23, 31, 40, 50, 56, 127, 134
Tompkins, Daniel D., 111

U
Union Metallic Cartridge & Cap Company, 47

V
Venditti, 20, 145
Vesuvius, Mt., 23
Volcanic, origin of the name, 22
Volition Repeater, 9–13, 15, 16, 43, 44
Volcanic Repeating Arms Company, 45, 47–48, 52–54, 57, 59, 62–67, 85–86, 88, 95, 96, 117–126
Volcanic Pistol Cartridges, see Cartridges

W
Walch Fire-Arms Company, 127
Wesson, Daniel B., 16, 19–23, 25, 41, 44, 46, 51, 52, 58, 66
Wesson, Edwin, 19
Whitney Arms Company, 52
Whitney, Eli, 16
Winchester, Oliver F., 12, 16, 32, 34, 45, 47, 51–53, 66–69, 83, 84, 95, 96, 116, 117, 127, 128, 130, 142
Winchester Repeating Arms Company, 16, 50, 52, 53, 83, 96, 99, 113, 131, 133, 134

Y
Yellow Fever War of 1858, 111–113
Young, Gustave, 95

about the authors...

Edmund Lewis, M.D., a retired radiologist, has been interested in Winchester rifles from the age of fifteen and has collected them in earnest for over thirty years. He has come to enjoy and appreciate the history and background stories of these firearms and has written about them in a variety of magazines and recently has had published *The Story of the Winchester 1 of 1000 and 1 of 100 Rifles*. Volcanic firearms have also long held his interest and led to the present book. His other interests have included hunting in Africa, which he has done and written about for many years. He holds memberships in The National Rifle Association, The Winchester Arms Collectors Association, the Winchester Club of America and several other firearms collectors groups. Ed is married to Catherine Adams-Lewis who provided extensive advise, support and editing assistance during the writing of this book. He has four children, Madeleine, Jennifer, Ed and Sam and a number of wonderful grandchildren.

✣ ✣ ✣

Stephen Rutter, M.D. has actively pursued his interest in antique firearms collecting following retirement a number of years ago. He initially limited his collecting to the Winchester Model 1873, and it was the history of the development of that famous firearm that led him to discover the most interesting series of events depicted in this book. Other interests include woodworking, hiking and until recently, endurance horse riding on numerous western trails. He is a Life Member of the NRA, the Winchester Arms Collectors Association and numerous state collector associations where he has exhibited well-received displays depicting the history and the arms he collects. He lives in a mountainous area of southern Arizona with his wife and eighteen “family members” that include horses, dogs and cats.